Principles of
Optimal
Control Theory

MATHEMATICAL CONCEPTS AND METHODS IN SCIENCE AND ENGINEERING

Series Editor: **Angelo Miele**
*Mechanical Engineering and Mathematical Sciences
Rice University, Houston, Texas*

Volume 1	**INTRODUCTION TO VECTORS AND TENSORS** Volume 1: Linear and Multilinear Algebra *Ray M. Bowen and C.-C. Wang*
Volume 2	**INTRODUCTION TO VECTORS AND TENSORS** Volume 2: Vector and Tensor Analysis *Ray M. Bowen and C.-C. Wang*
Volume 3	**MULTICRITERIA DECISION MAKING AND DIFFERENTIAL GAMES** *Edited by George Leitmann*
Volume 4	**ANALYTICAL DYNAMICS OF DISCRETE SYSTEMS** *Reinhardt M. Rosenberg*
Volume 5	**TOPOLOGY AND MAPS** *Taqdir Husain*
Volume 6	**REAL AND FUNCTIONAL ANALYSIS** *A. Mukherjea and K. Pothoven*
Volume 7	**PRINCIPLES OF OPTIMAL CONTROL THEORY** *R. V. Gamkrelidze*
Volume 8	**INTRODUCTION TO THE LAPLACE TRANSFORM** *Peter K. F. Kuhfittig*
Volume 9	**MATHEMATICAL LOGIC** An Introduction to Model Theory *A. H. Lightstone*
Volume 10	**SINGULAR OPTIMAL CONTROLS** *R. Gabasov and F. M. Kirillova*
Volume 11	**INTEGRAL TRANSFORMS IN SCIENCE AND ENGINEERING** *Kurt Bernardo Wolf*

A Continuation Order Plan is available for this series. A continuation order will bring delivery of each new volume immediately upon publication. Volumes are billed only upon actual shipment. For further information please contact the publisher.

14988

Principles of Optimal Control Theory

R. V. Gamkrelidze
Steklov Institute, Moscow, U.S.S.R.

Translated from Russian by
Karol Makowski
Los Angeles, California

Translation editor
Leonard D. Berkovitz
Purdue University, West Lafayette, Indiana

Plenum Press · New York and London

Library of Congress Cataloging in Publication Data

Gamkrelidze, R   V
  Principles of optimal control theory.

  (Mathematical concepts and methods in science and engineering)
  Translation of Osnovy optimal'nogo upravleniia.
  Based on lectures presented in 1974 at Tbilisi State University.
  1. Control theory. 2. Mathematical optimization. I. Title.
QA402.3.G3513   629.8'312   77-10742
ISBN 0-306-30977-7

The original Russian text, published by Tbilisi University Publishing House in Tbilisi in 1975, has been corrected by the author for the present edition. This translation is published under an agreement with the Copyright Agency of the USSR (VAAP).

ОСНОВЫ ОПТИМАЛЬНОГО УПРАВЛЕНИЯ

Р. В. ГАМКРЕЛИДЗЕ

OSNOVY OPTIMAL'NOGO UPRAVLENIYA

R. V. Gamkrelidze

© 1978 Plenum Press, New York
A Division of Plenum Publishing Corporation
227 West 17th Street, New York, N.Y. 10011

All rights reserved

No part of this book may be reproduced, stored in a retrieval system, or transmitted, in any form or by any means, electronic, mechanical, photocopying, microfilming, recording, or otherwise, without written permission from the Publisher

Printed in the United States of America

TO MY PARENTS

Foreword

In the late 1950's, the group of Soviet mathematicians consisting of L. S. Pontryagin, V. G. Boltyanskii, R. V. Gamkrelidze, and E. F. Mishchenko made fundamental contributions to optimal control theory. Much of their work was collected in their monograph, *The Mathematical Theory of Optimal Processes*. Subsequently, Professor Gamkrelidze made further important contributions to the theory of necessary conditions for problems of optimal control and general optimization problems. In the present monograph, Professor Gamkrelidze presents his current view of the fundamentals of optimal control theory. It is intended for use in a one-semester graduate course or advanced undergraduate course. We are now making these ideas available in English to all those interested in optimal control theory.

West Lafayette, Indiana, USA   Leonard D. Berkovitz
                                                                               Translation Editor

Preface

This book is based on lectures I gave at the Tbilisi State University during the fall of 1974. It contains, in essence, the principles of general control theory and proofs of the maximum principle and basic existence theorems of optimal control theory.

Although the proofs of the basic theorems presented here are far from being the shortest, I think they are fully justified from the conceptual viewpoint. In any case, the notions we introduce and the methods developed have one unquestionable advantage—they are constantly used throughout control theory, and not only for the proofs of the theorems presented in this book.

I restricted myself to the presentation of the time-optimal problem with fixed end points, since it is less cumbersome from the viewpoint of the formulation of results and, at the same time, it contains all the essential difficulties peculiar to optimal control problems. The transfer to problems with general boundary conditions and more general functionals can be accomplished, as a rule, without any particular difficulties and, in most cases, in an automatic way.

The material presented is intended for use in a one-semester course for those students and post-graduates specializing in optimization theory in applied mathematics departments. It is assumed that the students are already familiar with the elements of the theory of optimal control, including the principles of linear systems theory and convex set theory. It is also assumed that the reader has mastered the very basic notions of general measure theory.

In conclusion, I would like to express my sincere gratitude to my post-graduate student A. Agrachev for his significant help in preparing the initial notes of the lectures. His help was particularly useful in writing the final chapter (Chapter 8), which covers material not included in the course of lectures.

Tbilisi, Georgia, USSR R. Gamkrelidze

Contents

1. **Formulation of the Time-Optimal Problem and Maximum Principle** 1
   1.1. Statement of the Optimal Problem 1
   1.2. On the Canonical Systems of Equations Containing a Parameter and on the Pontryagin Maximum Condition ... 4
   1.3. The Pontryagin Maximum Principle 8
   1.4. A Geometrical Interpretation of the Maximum Condition . 11
   1.5. The Maximum Condition in the Autonomous Case 12
   1.6. The Case of an Open Set U. The Canonical Formalism for the Solution of Optimal Control Problems 16
   1.7. Concluding Remarks 19

2. **Generalized Controls** 21
   2.1. Generalized Controls and a Convex Control Problem 21
   2.2. Weak Convergence of Generalized Controls 28

3. **The Approximation Lemma** 37
   3.1. Partition of Unity 38
   3.2. The Approximation Lemma 45

4. **The Existence and Continuous Dependence Theorem for Solutions of Differential Equations** 53
   4.1. Preparatory Material 53
   4.2. A Fixed-Point Theorem for Contraction Mappings 60

	4.3.	The Existence and Continuous Dependence Theorem for Solutions of Equation (4.3)	63
	4.4.	The Spaces $E_{\text{Lip}}(G)$	69
	4.5.	The Existence and Continuous Dependence Theorems for Solutions of Differential Equations in the General Case	72
5.	**The Variation Formula for Solutions of Differential Equations**		**79**
	5.1.	The Spaces E_1 and $E_1(G)$	79
	5.2.	The Equation of Variation and the Variation Formula for the Solution	82
	5.3.	Proof of Theorem 5.1	87
	5.4.	A Counterexample	90
	5.5	On Solutions of Linear Matrix Differential Equations	93
6.	**The Varying of Trajectories in Convex Control Problems**		**99**
	6.1.	Variations of Generalized Controls and the Corresponding Variations of the Controlled Equation	99
	6.2.	Variations of Trajectories	107
7.	**Proof of the Maximum Principle**		**115**
	7.1.	The Integral Maximum Condition, the Pontryagin Maximum Condition, and Their Equivalence	115
	7.2.	The Maximum Principle in the Class of Generalized Controls	118
	7.3.	Construction of the Cone of Variations	120
	7.4.	Proof of the Maximum Principle	130
8.	**The Existence of Optimal Solutions**		**135**
	8.1.	The Weak Compactness of the Class of Generalized Controls	136
	8.2.	The Existence Theorem for Convex Optimal Problems	146
	8.3.	The Existence Theorem in the Class of Ordinary Controls	150
	8.4.	Sliding Optimal Regimes	156
	8.5.	The Existence Theorem for Regular Problems of the Calculus of Variations	162
Bibliography			**171**
Index			**173**

1

Formulation of the Time-Optimal Problem and Maximum Principle

1.1. Statement of the Optimal Problem

We consider the following differential equation in R^n,

$$\dot{x} = f(t, x, u), \qquad (1.1)$$

which we shall call the *controlled equation*. The point

$$x = \begin{pmatrix} x^1 \\ \vdots \\ x^n \end{pmatrix} \in R^n$$

will be called the *phase point*, the parameter (point)

$$u = \begin{pmatrix} u^1 \\ \vdots \\ u^r \end{pmatrix} \in R^r$$

will be called the *control parameter*, and the vector

$$f(t, x, u) = \begin{pmatrix} f^1(t, x, u) \\ \vdots \\ f^n(t, x, u) \end{pmatrix} \in R^n$$

will be called the *phase velocity vector*. We assume that f is a continuous function on

$$R^{1+n+r} = \{(t, x, u): t \in R, x \in R^n, u \in R^r\},$$

and that it has a continuous derivative with respect to x:

$$\frac{\partial f(t, x, u)}{\partial x} = \left(\frac{\partial f(t, x, u)}{\partial x^1}, \ldots, \frac{\partial f(t, x, u)}{\partial x^n} \right).$$

An arbitrary set

$$U \subset R^r$$

is given in the "space of parameters" R^r. This set is said to be the *set of admissible values* of the control parameter. An arbitrary measurable and bounded function $u(t)$ which is defined on R and takes on its values in the set U is said to be an *admissible control*. The set of all admissible controls Ω_U will be called the *class of admissible controls*.

If an arbitrary admissible control $u(t)$ is substituted for the parameter u in the controlled equation (1.1), then we obtain the differential equation

$$\dot{x} = f(t, x, u(t)) = F(t, x),$$

whose right-hand side $F(t, x)$ is continuously differentiable with respect to x and measurable with respect to t. Such equations are studied in detail in Chapters 4 and 5, where all the necessary facts concerning these equations are proven. Now we note only the following: Any absolutely continuous function

$$x(t), \qquad t_1 \leq t \leq t_2,$$

is said to be a *solution* of this equation on the interval $t_1 \leq t \leq t_2$ if it satisfies the equality

$$\dot{x}(t) = f(t, x(t), u(t)) = F(t, x(t))$$

for almost all $t \in [t_1, t_2]$. By virtue of the absolute continuity of $x(t)$, this is equivalent to the statement that $x(t)$ satisfies the integral equation

$$x(t) = x(t_1) + \int_{t_1}^{t} F(\theta, x(\theta))\, d\theta$$

for all $t \in [t_1, t_2]$.

Moreover, it is clear that if $u_1(t)$ and $u_2(t)$ coincide for almost all $t \in R$, then the differential equations

$$\dot{x} = f(t, x, u_1(t)) \quad \text{and} \quad \dot{x} = f(t, x, u_2(t))$$

are equivalent.

A *controlled object* is given by a controlled equation (1.1) and a class of admissible controls Ω_U. The "state" of such an object is described by the *phase point* x, and its *phase motion* can be controlled by the choice of an admissible control $u(t) \in \Omega_U$. If, at the instant of time $t = t_1$, the *initial state* x_1 of the object is given and the control $u(t)$ is chosen, then the phase motion of the object is determined uniquely as the solution of the equation

$$\dot{x} = f(t, x, u(t)).$$

We shall identify the controlled object with the *control problem* which is given by the controlled equation (1.1) and the class of admissible controls Ω_U. We shall write the control problem in the form

$$\dot{x} = f(t, x, u), \quad u(t) \in \Omega_U. \quad (1.2)$$

The time-optimal problem with fixed end points is formulated as follows:

Two points x_1 and x_2 in the phase space R^n and an initial time t_1 are given. We are to choose an admissible control

$$\tilde{u}(t) \in \Omega_U$$

in such a way that the differential equation

$$\dot{x} = f(t, x, \tilde{u}(t))$$

has a solution $\tilde{x}(t)$ defined on the time interval $t_1 \leq t \leq t_2$, satisfying the boundary conditions

$$\tilde{x}(t_1) = x_1, \quad \tilde{x}(t_2) = x_2,$$

and such that the time of transfer from x_1 to x_2 is minimal:

$$t_2 - t_1 \to \min.$$

Thus, if, for any other admissible control $\hat{u}(t)$, there exists a solution $\hat{x}(t)$, $t_1 \leq t \leq \hat{t}_2$, of the equation

$$\dot{x} = f(t, x, \hat{u}(t))$$

that satisfies the same boundary conditions

$$\hat{x}(t_1) = x_1, \quad \hat{x}(\hat{t}_2) = x_2,$$

then $\hat{t}_2 - t_1 \geq t_2 - t_1$, i.e., $\hat{t}_2 \geq t_2$.

The control $\tilde{u}(t)$ is called a *(time) optimal control*, $\tilde{x}(t)$ is called the corresponding *optimal trajectory*, and $t_2 - t_1$ is called the *optimal transfer time* from x_1 to x_2 with the initial condition $x(t_1) = x_1$.

The pair $\tilde{u}(t), \tilde{x}(t)$ with $t_1 \leq t \leq t_2$ will also be called an optimal solution, or a solution of the *(time) optimal problem*

$$\begin{aligned} &\dot{x} = f(t, x, u), \quad u(t) \in \Omega_U, \quad t = t_1, \\ &x(t_1) = x_1, \quad x(t_2) = x_2, \quad t_2 - t_1 \to \min. \end{aligned} \quad (1.3)$$

It should be noted that the values of t_1, $x(t_1)$, and $x(t_2)$ are given in the boundary conditions of the problem, but the value of t_2 is not. In the case of an *autonomous* equation (1.1), i.e., when the right-hand side does not depend on t, the fact that the initial time t_1 is fixed in the boundary conditions obviously has no significance.

The optimal problem can also be studied for controlled objects which are determined by the same controlled equation (1.1), but by a class of admissible controls wider than Ω_U. We shall, however restrict ourselves to the definition already given.

Sometimes it is useful to narrow down the class Ω_U by considering, e.g., only *piecewise-continuous* or even only *piecewise-constant* controls which take on their values in U. In this connection, a control $u(t) \in \Omega_U$ is said to be piecewise-continuous if it has a finite number of points of discontinuity on any bounded interval of the t-axis and if the right- and left-hand limits exist at every point of discontinuity. A control $u(t) \in \Omega_U$ is said to be piecewise-constant if the t-axis can be partitioned into a finite number of intervals on each of which $u(t)$ has a constant value.

1.2. On the Canonical Systems of Equations Containing a Parameter and on the Pontryagin Maximum Condition

First, we shall establish the notation. Column vectors will, as before, be denoted by roman letters, while row vectors will always be denoted by Greek letters, e.g., $\xi = (\xi_1, \ldots, \xi_n)$. The scalar product of a row vector and a column vector of the same dimensionality will be written in the form of matrix multiplication:

$$\xi \cdot x = (\xi_1, \ldots, \xi_n) \begin{pmatrix} x^1 \\ \vdots \\ x^n \end{pmatrix} = \sum_{i=1}^{n} \xi_i x^i.$$

The Jacobian of a vector-valued function $f(t, x, u)$ with respect to the coordinates of the vector x will be denoted by

$$f_x(t, x, u) = \frac{\partial f(t, x, u)}{\partial x} = \left(\frac{\partial f^i(t, x, u)}{\partial x^j} \right), \quad i, j = 1, \ldots, n.$$

By the absolute value of an arbitrary $n \times m$ matrix $A = (a_{ij})$, $i = 1, \ldots, n$,

Time-Optimal Problem and Maximum Principle

$j = 1, \ldots, m$, we will mean the Euclidean norm

$$|A| = \left(\sum_{i,j} a_{ij}^2 \right)^{1/2}.$$

We define a scalar-valued function of the time t and three vector arguments

$$x = \begin{pmatrix} x^1 \\ \vdots \\ x^n \end{pmatrix}, \quad \psi = (\psi_1, \ldots, \psi_n), \quad u = \begin{pmatrix} u^1 \\ \vdots \\ u^r \end{pmatrix}$$

by the formula

$$H(t, x, \psi, u) = \psi f(t, x, u) = \sum_{i=1}^{n} \psi_i f^i(t, x, u).$$

The function H is a scalar product of the vector ψ and the phase velocity vector $f(t, x, u)$ of the controlled object. This function is linear in ψ and continuously differentiable with respect to x.

We use the notation

$$\frac{\partial H}{\partial \psi} = \begin{pmatrix} \dfrac{\partial H}{\partial \psi_1} \\ \vdots \\ \dfrac{\partial H}{\partial \psi_n} \end{pmatrix} = \begin{pmatrix} f^1 \\ \vdots \\ f^n \end{pmatrix} = f,$$

$$\frac{\partial H}{\partial x} = \left(\frac{\partial H}{\partial x^1}, \ldots, \frac{\partial H}{\partial x^n} \right) = \left(\sum_{i=1}^{n} \psi_i f^i_{x^1}(t, x, u), \ldots, \sum_{i=1}^{n} \psi_i f^i_{x^n}(t, x, u) \right)$$

$$= \psi f_x(t, x, u).$$

We now write the following system of two n-dimensional vector differential equations for vector-valued functions x and ψ:

$$\dot{x} = \frac{\partial H(t, x, \psi, u)}{\partial \psi} = f(t, x, u),$$

$$\dot{\psi} = -\frac{\partial H(t, x, \psi, u)}{\partial x} = -\psi f_x(t, x, u).$$
(1.4)

This system is called the *Hamiltonian*, or *canonical system* of differential equations in x and ψ. The function H is called the *Hamiltonian* of this system. Besides the variables x and ψ, the function H also contains the time t and the argument u, which we shall consider as a parameter.

In accordance with commonly accepted terminology, every scalar-valued, twice continuously differentiable function $\mathcal{H}(t, x, \psi)$ generates the corresponding *Hamiltonian* or *canonical system*

$$\dot{x} = \frac{\partial \mathcal{H}(t, x, \psi)}{\partial \psi} = \begin{pmatrix} \frac{\partial \mathcal{H}}{\partial \psi_1} \\ \vdots \\ \frac{\partial \mathcal{H}}{\partial \psi_n} \end{pmatrix},$$

$$\dot{\psi} = -\frac{\partial \mathcal{H}(t, x, \psi)}{\partial x} = -\left(\frac{\partial \mathcal{H}}{\partial x^1}, \ldots, \frac{\partial \mathcal{H}}{\partial x^n}\right).$$

(1.5)

The function $\mathcal{H}$ is said to be the *Hamiltonian* of this system.

A remarkable property of this system is the fact that, if $x(t)$, $\psi(t)$ with $t_1 \leq t \leq t_2$ is an arbitrary solution of the system, then

$$\frac{d}{dt} \mathcal{H}(t, x(t), \psi(t)) = \frac{\partial}{\partial t} \mathcal{H}(t, x(t), \psi(t)),$$

where on the left-hand side we have the total derivative with respect to t, and on the right-hand side—the partial derivative of $\mathcal{H}(t, x, \psi)$ with respect to t. In particular, if $\mathcal{H}$ does not depend explicitly on t, then

$$\frac{d}{dt} \mathcal{H}(x(t), \psi(t)) = 0.$$

The proof consists of the direct computation:

$$\frac{d}{dt} \mathcal{H}(t, x(t), \psi(t)) = \dot{\psi} \frac{\partial \mathcal{H}}{\partial \psi} + \frac{\partial \mathcal{H}}{\partial x} \dot{x} + \frac{\partial \mathcal{H}}{\partial t}$$

$$= -\frac{\partial \mathcal{H}}{\partial x} \frac{\partial \mathcal{H}}{\partial \psi} + \frac{\partial \mathcal{H}}{\partial x} \frac{\partial \mathcal{H}}{\partial \psi} + \frac{\partial \mathcal{H}}{\partial t} = \frac{\partial \mathcal{H}}{\partial t}.$$

A solution $x(t)$, $\psi(t)$ of the system (1.5), which does not contain any parameter, is uniquely determined by the initial conditions

$$t = t_1, \quad x(t_1) = \hat{x}, \quad \psi(t_1) = \hat{\psi}.$$

In contrast, a solution of the system (1.4) is determined uniquely only if we substitute in H a particular function of t for the parameter u. So far we are not constrained in our choice of such a function, except for the condition $u(t) \in \Omega_U$.

In order to make a definite choice, we assume that we are given a relation which connects t, x, ψ, and u, and which we write as

$$\omega(t, x, \psi, u) = 0. \tag{1.6}$$

In principle, this relation should allow us to express u in terms of t, x, and ψ (for the time being, we do not specify the form of this function in any way). In so doing, we eliminate the parameter u from equations (1.4). Hence, we find

$$u = u(t, x, \psi),$$

and substituting this expression for u in (1.4), we obtain the following system of differential equations in x, ψ:

$$\dot{x} = f(t, x, u(t, x, \psi)), \qquad \dot{\psi} = -\psi f_x(t, x, u(t, x, \psi)).$$

With initial conditions given, this system is uniquely solvable under the assumption that the function $u(t, x, \psi)$ is sufficiently regular, e.g., continuously differentiable.

The condition for the regular solvability of equation (1.6) in u is quite restrictive. Therefore, we replace the procedure of eliminating u from (1.4) with the aid of (1.6) by some "substitute" procedure. To this end, we give the following precise definition:

Suppose there exist absolutely continuous functions $x(t)$ and $\psi(t)$, $t_1 \leq t \leq t_2$, and a measurable function $u(t) \in \Omega_U$ that satisfy the system of differential equations (1.4), equation (1.6), and the initial conditions

$$t = t_1, \qquad x(t_1) = x_1, \qquad \text{and} \qquad \psi(t_1) = \psi_1.$$

Then we shall say that the system of functions $x(t)$, $\psi(t)$, and $u(t)$ is the solution of the canonical system (1.4) with given initial conditions, and that this solution is obtained by the *elimination of the parameter u* from the canonical system (1.4) with the aid of relation (1.6). Since the function $u(t)$ is measurable, it is sufficient to assume that the equality

$$\omega(x(t), \psi(t), u(t), t) = 0$$

holds for almost all $t \in [t_1, t_2]$.

The case of the unique and regular solvability of equation (1.6) for u is encountered in the so-called regular problem of the classical calculus of variations. We shall consider it at the end of this chapter.

Our main interest lies in the following form of condition (1.6), which is called the Pontryagin maximum condition:

We denote by $M(t, x, \psi)$ the function

$$M(t, x, \psi) = \sup_{u \in U} H(t, x, \psi, u).$$

The condition
$$H(t, x, \psi, u) = M(t, x, \psi),$$
which serves to eliminate the parameter u from the canonical system (1.4), will be called the *Pontryagin maximum condition*.

If $x(t)$, $\psi(t)$, $u(t)$, $t_1 \leq t \leq t_2$, form a solution of the canonical system (1.4) and satisfy the Pontryagin maximum condition for almost all $t \in [t_1, t_2]$, i.e., the equality
$$H(t, x(t), \psi(t), u(t)) = M(t, x(t), \psi(t)),$$
then we shall say that the solution under consideration was obtained by the *elimination of the parameter u from the canonical system (1.4) with the aid of the Pontryagin maximum condition*.

1.3. The Pontryagin Maximum Principle

We shall now formulate Theorem 1.1, i.e., the Pontryagin maximum principle, which is a necessary condition satisfied by any solution of the optimal problem (1.3), and we shall discuss the possibilities of using this condition to actually find optimal solutions. The proof of the maximum principle will be given in Chapter 7.

Theorem 1.1. Let
$$\tilde{u}(t) \in \Omega_U, \qquad \tilde{x}(t), \qquad t_1 \leq t < t_2,$$
be a solution of the optimal problem (1.3). Then there exists a nonzero, absolutely continuous function
$$\tilde{\psi}(t) = (\tilde{\psi}_1(t), \ldots, \tilde{\psi}_n(t)), \qquad t_1 \leq t \leq t_2,$$
such that for almost all t on the interval $t_1 \leq t \leq t_2$, the system of functions $\tilde{u}(t)$, $\tilde{x}(t)$, and $\tilde{\psi}(t)$ satisfies the canonical system
$$\dot{x} = \frac{\partial}{\partial \psi} H(t, x, \psi, u), \qquad \dot{\psi} = -\frac{\partial}{\partial x} H(t, x, \psi, u), \tag{1.7}$$
with the Hamiltonian
$$H(t, x, \psi, u) = \psi f(t, x, u), \tag{1.8}$$
and the Pontryagin maximum condition
$$H(t, \tilde{x}(t), \tilde{\psi}(t), \tilde{u}(t)) = M(t, \tilde{x}(t), \tilde{\psi}(t)). \tag{1.9}$$

Moreover, the function $M(t, \tilde{x}(t), \tilde{\psi}(t))$, $t_1 \leqslant t \leqslant t_2$, is a continuous function of t, and the following inequality holds at $t = t_2$:

$$M(t_2, \tilde{x}(t_2), \tilde{\psi}(t_2)) \geqslant 0. \tag{1.10}$$

Both the maximum condition (1.9) and inequality (1.10) can be given a simple and geometrically descriptive interpretation, which we shall discuss in the next section.

Any solution $u(t)$, $x(t)$, $\psi(t)$ of the system (1.7)–(1.10) with a nonzero $\psi(t)$ that satisfies the boundary conditions

$$t = t_1, \qquad x(t_1) = x_1, \qquad x(t_2) = x_2 \tag{1.11}$$

will be called an *extremal solution* or *an extremal* of the optimal problem (1.3). The function

$$H(t, x, \psi, u) = \psi f(t, x, u)$$

will be called the *Hamiltonian* of the optimal problem (1.3).

The maximum principle asserts that any solution of the optimal problem (1.3) is contained among the extremals of this problem. Therefore, we shall briefly discuss the following problem: To what extent do the boundary conditions (1.11), together with the maximum condition (1.9) and inequality (1.10), determine the solution of the canonical system (1.7)? In this connection, the maximum condition should be considered as a relation of the form (1.6) between t, x, ψ, and u, which eliminates the parameter u from the system (1.7).

First of all, it is clear that the assumption that the function $\tilde{\psi}(t)$ does not vanish identically is essential. Because of the linearity of H in ψ, any solution $u(t)$, $x(t)$ of the system

$$\dot{x} = \frac{\partial}{\partial \psi} H(t, x, \psi, u) = f(t, x, u), \qquad u(t) \in \Omega_U,$$
$$t = t_1, \qquad x(t_1) = x_1, \qquad x(t_2) = x_2$$

together with the function $\psi(t) \equiv 0$ satisfies the system (1.7)–(1.11). Therefore, without the condition $\tilde{\psi}(t) \not\equiv 0$, the maximum principle would become an empty assertion. With a nonzero $\psi(t)$, the maximum principle, "generally speaking," eliminates the parameter u from system (1.7) in a satisfactory manner, and we obtain one or several isolated solutions of the boundary problem (1.7)–(1.11).

It is important to note that, if $\tilde{\psi}(t) \not\equiv 0$, then $\tilde{\psi}(t) \neq 0$ $\forall t \in [t_1, t_2]$, since

$\tilde{\psi}(t)$ is a solution of the linear homogeneous equation

$$\dot{\psi} = -\psi f_x(t, \tilde{x}(t), \tilde{u}(t)).$$

The fact that there are no boundary conditions imposed on $\psi(t)$ is entirely natural. By a simple count of the "number of essential parameters," one can easily convince oneself that the number of "essential parameters" of the system (1.7)–(1.11) is equal to the number of its "independent conditions" and, therefore, the system is not overdetermined. However, an addition of any new (independent) condition will lead to an overdetermined system.

Indeed, let us take the initial values t_1, $x(t_1)$, $\psi(t_1)$, and the final time t_2 for the parameters of the boundary problem (1.7)–(1.11), i.e., a total of $2n+2$ scalar parameters. If t_1, $x(t_1)$, and $\psi(t_1)$ are given, then the solution $u(t), x(t), \psi(t)$ of the system (1.7)–(1.9) will evolve in time in a unique way, under the assumption that the maximum condition eliminates u from (1.7), i.e., that the equation

$$H(t, x(t), \psi(t), u(t)) = M(t, x(t), \psi(t))$$

is uniquely solvable in u for almost all t. Since t_1 and $x(t_1)$ are fixed in the boundary condition (1.11), $n+1$ scalar parameters $\psi(t_1)$, t_2 remain free. Using these parameters, we are to satisfy the remaining n scalar conditions $x(t_2) = x_2$.

We shall show that, if

$$u(t), x(t), \psi(t), \qquad t_1 \leq t \leq t_2,$$

is a solution of the system (1.7)–(1.11), then the system of functions

$$u(t), x(t), \lambda\psi(t), \qquad t_1 \leq t \leq t_2,$$

where λ is an arbitrary nonnegative constant, is also a solution of the system (1.7)–(1.11). Therefore, there are in fact only $n-1$ essential parameters among the n coordinates of the vector $\psi(t_1)$ and, counting t_2, we have at our disposal only n essential parameters, with the aid of which we must satisfy the boundary condition $x(t_2) = x_2$ at the right end point.

The assertion becomes obvious if we note that the function H is linear in ψ, the function $M(t, x, \psi)$ is positively homogeneous in ψ,

$$M(t, x, \lambda\psi) = \lambda M(t, x, \psi) \qquad \forall \lambda \geq 0,$$

the left-hand side of inequality (1.10) is greater than zero, and finally, that the boundary conditions (1.11) do not contain any values of $\psi(t)$.

Since the solutions of the optimal problem (1.3) will necessarily be among

Time-Optimal Problem and Maximum Principle

the solutions of the boundary value problem (1.7)–(1.11), the methods for solving such boundary value problems are of great importance, and special chapters in optimization theory are devoted to them.*

1.4. A Geometrical Interpretation of the Maximum Condition

We denote by $P(t, x)$ the set of all possible phase velocities in the control problem (1.2) for given values of t and x:

$$P(t, x) = \{f(t, x, u): u \in U\} \subset R^n.$$

Consider an extremal $\tilde{u}(t)$, $\tilde{x}(t)$, $\tilde{\psi}(t)$, $t_1 \leq t \leq t_2$, of problem (1.3), and let $\Pi_t \subset R^n$ be the $(n-1)$-dimensional subspace which is orthogonal to $\tilde{\psi}(t)$ at the time t. The maximum condition (1.9) asserts that, for almost all $t \in [t_1, t_2]$, the scalar product of the vector $\tilde{\psi}(t)$ with the phase velocity vector $f(t, \tilde{x}(t), \tilde{u}(t))$ (of the extremal trajectory under consideration) majorizes the scalar product of $\tilde{\psi}(t)$ with any possible phase velocity of the controlled object (1.2):

$$\tilde{\psi}(t) f(t, \tilde{x}(t), \tilde{u}(t)) \geq \tilde{\psi}(t) p \qquad \forall p \in P(t, \tilde{x}(t)).$$

Therefore, the set $P(t, \tilde{x}(t))$ will lie, for almost all $t \in [t_1, t_2]$, in one of the two closed half-spaces into which the hyperplane

$$f(t, \tilde{x}(t), \tilde{u}(t)) + \Pi_t = \{f(t, \tilde{x}(t), \tilde{u}(t)) + p: p \in \Pi_t\}$$

subdivides the space R^n. Thus, this hyperplane is a support hyperplane of the set $P(t, \tilde{x}(t))$, and the point $f(t, \tilde{x}(t), \tilde{u}(t))$ is a boundary point of $P(t, \tilde{x}(t))$.

One can visualize the "kinematic" picture of the extremal motion by assuming that, for almost all $t \in [t_1, t_2]$, the controlled object chooses the phase velocity $f(t, \tilde{x}(t), \tilde{u}(t))$ in such a way as to move in the direction of the vector $\tilde{\psi}(t)$ with the greatest possible speed.

According to (1.10), it is always possible to choose a $\hat{u} \in U$ at the final

---

*In essence, the optimal problem (1.3) is "infinite-dimensional" in the sense that an optimal control is chosen from an "infinite-dimensional" family of functions Ω_U. The fact that "constraints" in the form of the differential equation and boundary conditions

$$\dot{x} = f(t, x, u), \qquad t = t_1, \qquad x(t_1) = x_1, \qquad x(t_2) = x_2$$

are imposed on controls $u(t) \in \Omega_U$ does not, generally speaking, make finite the "dimensionality" of the family of all admissible controls that satisfies these constraints.

By taking into account the optimality of the control and using the maximum principle, we reduce the initial "infinite-dimensional" problem to a complicated, but nevertheless finite-dimensional boundary value problem (1.7)–(1.11).

time $t=t_2$ such that the projection of the phase velocity $f(t_2, \tilde{x}(t_2), \hat{u})$ onto the direction $\tilde{\psi}(t_2)$ is nonnegative. It is enough to take for $\hat{u}$ a point at which the supremum

$$\sup_{u \in U} \tilde{\psi}(t_2) f(t_2, \tilde{x}(t_2), u) = M(t_2, \tilde{x}(t_2), \tilde{\psi}(t_2)) \geq 0$$

is attained. If the supremum is not attained, then we can only guarantee that the absolute value of the scalar product $\tilde{\psi}(t_2) f(t_2, \tilde{x}(t_2), u)$, $u \in U$, can be made arbitrarily close to the supremum.

1.5. The Maximum Condition in the Autonomous Case

If the optimal problem (1.3) is autonomous, i.e., if the function $f(t, x, u)$ does not depend on t, then the value of $M(x, \psi)$ evaluated along an arbitrary extremal solution of problem (1.3) turns out to be constant and, on the basis of (1.10), nonnegative. Because this fact often has important applications, we shall formulate and prove it in the form of the following assertion:

Assertion 1.1. Let the system of function $\tilde{u}(t)$, $\tilde{x}(t)$, $\tilde{\psi}(t)$, $t_1 \leq t \leq t_2$, satisfy: (i) the canonical system

$$\dot{x} = \frac{\partial}{\partial \psi} H(x, \psi, u),$$

$$\dot{\psi} = -\frac{\partial}{\partial x} H(x, \psi, u),$$

where $H(x, \psi, u) = \psi f(x, u)$, (ii) the maximum condition

$$H(\tilde{x}(t), \tilde{\psi}(t), \tilde{u}(t)) = \sup_{u \in U} H(\tilde{x}(t), \tilde{\psi}(t), u) = M(\tilde{x}(t), \tilde{\psi}(t)) \quad (1.12)$$

for almost all $t \in [t_1, t_2]$, and (iii) the inequality

$$M(\tilde{x}(t_2), \tilde{\psi}(t_2)) \geq 0.$$

Then $M(\tilde{x}(t), \tilde{\psi}(t))$, as a function of t, is identically equal to a nonnegative constant:

$$M(\tilde{x}(t), \tilde{\psi}(t)) \equiv \text{const} \geq 0, \qquad t_1 \leq t \leq t_2.$$

Proof. We shall make use of the continuity of the function $M(\tilde{x}(t), \tilde{\psi}(t))$, $t_1 \leq t \leq t_2$, which will be proved in Chapter 7.

Time-Optimal Problem and Maximum Principle

By definition, the set of values of the control $\tilde{u}(t)$ is bounded, i.e., the set

$$N = \{\tilde{u}(t): t \in R\} \subset U \subset R^r$$

is bounded and, therefore, its closure $\overline{N}$ is compact. We define a function $v(t)$, $t_1 \leq t \leq t_2$, by the following condition: If, at a point t, $H(\tilde{x}(t), \tilde{\psi}(t), \tilde{u}(t)) = M(\tilde{x}(t), \tilde{\psi}(t))$, then $v(t) = \tilde{u}(t)$; in the opposite case $v(t)$ coincides with an arbitrary point $\hat{v} \in \overline{N}$ at which

$$H(\tilde{x}(t), \tilde{\psi}(t), \hat{v}) = \max_{u \in \overline{N}} H(\tilde{x}(t), \tilde{\psi}(t), u).$$

We use the notation

$$\hat{H}(t) = H(\tilde{x}(t), \tilde{\psi}(t), v(t)), \qquad t_1 \leq t \leq t_2.$$

We shall show that, for any $t \in [t_1, t_2]$,

$$\hat{H}(t) = \sup_{u \in \overline{U}} H(\tilde{x}(t), \tilde{\psi}(t), u), \tag{1.13}$$

where $\overline{U}$ is the closure of $U \subset R^r$. This equality is obvious for those t for which

$$\hat{H}(t) = M(\tilde{x}(t), \tilde{\psi}(t)) = \sup_{u \in U} H(\tilde{x}(t), \tilde{\psi}(t), u) = \sup_{u \in \overline{U}} H(\tilde{x}(t), \tilde{\psi}(t), u).$$

If we assume that

$$\hat{H}(t) = \max_{u \in \overline{N}} H(\tilde{x}(t), \tilde{\psi}(t), u) < \sup_{u \in \overline{U}} H(\tilde{x}(t), \tilde{\psi}(t)u),$$

then there exists a point $u' \in U$ at which

$$\hat{H}(t) = \max_{u \in \overline{N}} H(\tilde{x}(t), \tilde{\psi}(t), u) < H(\tilde{x}(t), \tilde{\psi}(t), u').$$

Since the set $\overline{N}$ is compact, and since $\tilde{x}(t)$, $\tilde{\psi}(t)$, and $H(x, \psi, u)$ are continuous functions of their arguments, we obtain the following inequality for all points τ that are sufficiently close to the point t and for any point $u \in \overline{N}$,

$$H(\tilde{x}(\tau), \tilde{\psi}(\tau), u) < H(\tilde{x}(\tau), \tilde{\psi}(\tau), u'),$$

which contradicts the maximum condition (1.12).

The equality $M(\tilde{x}(t), \tilde{\psi}(t)) = \text{const} \geq 0$ will be proved if we show that the function $\hat{H}(t)$, $t_1 \leq t \leq t_2$, is absolutely continuous and that its derivative vanishes almost everywhere. Indeed, this will prove that $\hat{H}(t) = \text{const}$, and we known by definition that the continuous function $M(\tilde{x}(t), \tilde{\psi}(t)) = \hat{H}(t)$ for almost all $t \in [t_1, t_2]$.

Using the notation
$$\hat{H}(t', t'') = H(\tilde{x}(t'), \tilde{\psi}(t'), v(t'')),$$
we obtain, on the basis of (1.13),
$$\hat{H}(t'', t') - \hat{H}(t') \leq \hat{H}(t'') - \hat{H}(t'') \leq \hat{H}(t'') - \hat{H}(t', t'') \qquad \forall t', t'' \in [t_1, t_2].$$

The function $\hat{H}(t, \tau)$ is absolutely continuous in its first argument (for a fixed τ). Therefore,
$$\int_{t'}^{t''} \left(\frac{d\hat{H}(t, t')}{dt}\right) dt \leq \hat{H}(t'') - \hat{H}(t') \leq \int_{t'}^{t''} \left(\frac{d\hat{H}(t, t'')}{dt}\right) dt \qquad \forall t', t'' \in [t_1, t_2], t' \leq t''.$$

Hence
$$-\int_{t'}^{t''} \left|\frac{d\hat{H}(t, t')}{dt}\right| dt \leq \hat{H}(t'') - \hat{H}(t') \leq \int_{t'}^{t''} \left|\frac{d\hat{H}(t, t'')}{dt}\right| dt \qquad \forall t', t'' \in [t_1, t_2], t' \leq t''.$$
(1.14)

Since the functions $\tilde{u}(t)$ and $v(t)$ with $t_1 \leq t \leq t_2$ are bounded, we have for almost all $t \in [t_1, t_2]$
$$\left|\frac{d\hat{H}(t, \tau)}{dt}\right| = |-\psi(t)f_x(\tilde{x}(t), \tilde{u}(t))f(\tilde{x}(t), v(\tau)) + \tilde{\psi}(t)f_x(\tilde{x}(t), v(\tau))f(\tilde{x}(t), \tilde{u}(t))| \leq C$$
$$\forall t, \tau \in [t_1, t_2].$$

Hence, taking into account (1.14), we obtain the estimate
$$|\hat{H}(t'') - \hat{H}(t')| \leq C|t'' - t'| \qquad \forall t', t'' \in [t_1, t_2].$$

The absolute continuity of $\hat{H}(t)$ follows from this estimate.

In order to prove the equality
$$\frac{d\hat{H}(t)}{dt} = 0$$
for almost all $t \in [t_1, t_2]$, we shall write the following obvious estimate, which holds for almost all $t \in [t_1, t_2]$:
$$\left|\frac{d\hat{H}(t, \tau)}{dt}\right| \leq |\tilde{\psi}(t)|\{|f_x(\tilde{x}(t), v(\tau)) - f_x(\tilde{x}(t), \tilde{u}(t))| \cdot |f(\tilde{x}(t), \tilde{u}(t))|$$
$$+ |f(\tilde{x}(t), \tilde{u}(t)) - f(\tilde{x}(t), v(\tau))| \cdot |f_x(\tilde{x}(t), \tilde{u}(t))|\}$$
$$\leq D\{|f_x(\tilde{x}(t), \tilde{u}(t)) - f_x(\tilde{x}(t), v(\tau))| + |f(\tilde{x}(t), \tilde{u}(t)) - f(\tilde{x}(t), v(\tau))|\}.$$

Time-Optimal Problem and Maximum Principle

We use the notation

$$g(t, \tau) = |f_x(\tilde{x}(t), \tilde{u}(t)) - f_x(\tilde{x}(t), v(\tau))| + |f(\tilde{x}(t), \tilde{u}(t)) - f(\tilde{x}(t), v(\tau))|.$$

Then the last estimate together with the estimate (1.14) yield

$$-D \int_{t'}^{t''} g(t, t') dt \leq \hat{H}(t'') - \hat{H}(t') \leq D \int_{t'}^{t''} g(t, t'') dt \qquad \forall t', t'' \in [t_1, t_2], \, t' \leq t''. \tag{1.15}$$

We assume that $\theta \in (t_1, t_2)$ is a Lebesgue point of the functions

$$\hat{H}(t), \quad f_x(\tilde{x}(t), \tilde{u}(t)), \quad f(\tilde{x}(t), \tilde{u}(t)), \qquad t_1 \leq t \leq t_2, \tag{1.16}$$

and that $\tilde{u}(\theta) = v(\theta)$. Such points form a set of full measure on the interval $t_1 < t < t_2$.

It is easy to see that the following relations hold at an arbitrary point θ:

$$\frac{1}{h} \int_{\theta}^{\theta+h} g(t, \theta) dt \to 0, \qquad \frac{1}{h} \int_{\theta-h}^{\theta} g(t, \theta) dt \to 0 \qquad (h \to 0).$$

Indeed, since $\tilde{u}(\theta) = v(\theta)$, we have

$$g(t, \theta) \leq |f_x(\tilde{x}(t), \tilde{u}(t)) - f_x(\tilde{x}(\theta), \tilde{u}(\theta))| + |f(\tilde{x}(t), \tilde{u}(t)) - f(\tilde{x}(\theta), \tilde{u}(\theta))|$$
$$+ |f_x(\tilde{x}(t), \tilde{u}(\theta)) - f_x(\tilde{x}(\theta), \tilde{u}(\theta))| + |f(\tilde{x}(t), \tilde{u}(\theta)) - f(\tilde{x}(\theta), \tilde{u}(\theta))|,$$

The relations being proved now follow from the fact that θ is a Lebesgue point of the functions (1.16), and that the functions $f_x(\tilde{x}(t), \tilde{u}(\theta))$ and $f(\tilde{x}(t), \tilde{u}(\theta))$ as functions of t are continuous at the point $t = \theta$ for every fixed θ. Therefore, setting $t' = \theta$ and $t'' = \theta + h$ in the left estimate (1.15), dividing by $h > 0$, and letting h tend to zero, we obtain

$$0 \leq \frac{d\hat{H}(\theta)}{dt}$$

by virtue of the differentiability of the function $\hat{H}(t)$ at a Lebesgue point θ of this function. Similarly, the right estimate (1.15) yields

$$\frac{d\hat{H}(\theta)}{dt} \leq 0.$$

Therefore, $d\hat{H}(t)/dt = 0$ almost everywhere.

1.6. The Case of an Open Set U. The Canonical Formalism for the Solution of Optimal Control Problems

If the function $f(t, x, u)$ in the optimal problem (1.3) is twice continuously differentiable with respect to x and u, and if the set U of admissible values of the control is open in R^r, then a number of important corollaries can be derived from the maximum principle. These corollaries are well known in the classical calculus of variations.

First of all, it is clear that the maximum condition yields for almost all $t \in [t_1, t_2]$,

$$\frac{\partial}{\partial u^i} \hat{H}(t, \tilde{x}(t), \tilde{\psi}(t), \tilde{u}(t)) = 0, \qquad i = 1, \ldots, r,$$

because $H(t, \tilde{x}(t), \tilde{\psi}(t), u)$, as a function of u defined on an open set $U \subset R^r$, attains its maximum at the point $\tilde{u}(t)$. We write this system in the form of a single equation:

$$\frac{\partial}{\partial u} H(t, \tilde{x}(t), \tilde{\psi}(t), \tilde{u}(t)) = 0.$$

It follows also from the maximum condition that the $r \times r$ matrix

$$\left(\frac{\partial^2}{\partial u^i \partial u^j} H(t, \tilde{x}(t), \tilde{\psi}(t), \tilde{u}(t)) \right), \qquad i, j = 1, \ldots, r,$$

is not positive, i.e., the quadratic form

$$\sum_{i,j=1}^{r} \frac{\partial^2}{\partial u^i \partial u^j} H(t, \tilde{x}(t), \tilde{\psi}(t), \tilde{u}(t)) v^i v^j \leq 0$$

for arbitrary values $v^1, \ldots, v^r$. If the determinant of this matrix satisfies the additional condition

$$\left| \det \left(\frac{\partial^2}{\partial u^i \partial u^j} H(t, \tilde{x}(t), \tilde{\psi}(t), \tilde{u}(t)) \right) \right| \geq \alpha > 0$$

for almost all $t \in [t_1, t_2]$, then the quadratic form is negative-definite:

$$\sum_{i,j=1}^{r} \frac{\partial^2}{\partial u^i \partial u^j} H(t, \tilde{x}(t), \tilde{\psi}(t), \tilde{u}(t)) v^i v^j \leq -\beta |v|^2, \qquad \beta > 0.$$

In this case, we say that the extremal $\tilde{x}(t), \tilde{\psi}(t), \tilde{u}(t)$ with $t_1 \leq t \leq t_2$ is a *regular extremal* of the optimal problem (1.3).

The matrix
$$\left(\frac{\partial^2 H}{\partial u^i \partial u^j}\right)$$
can be considered as the Jacobian of the system of scalar-valued functions
$$\frac{\partial H}{\partial u} = \left(\frac{\partial H}{\partial u^1}, \ldots, \frac{\partial H}{\partial u^2}\right).$$
Thus, by virtue of the regularity condition
$$\left|\det\left(\frac{\partial^2 H}{\partial u^i \partial u^j}\right)\right| \geq \alpha > 0$$
and by the boundness of the set $\{\tilde{u}(t) : t \in [t_1, t_2]\}$, there exists a number $\varepsilon > 0$ that does not depend on $t \in [t_1, t_2]$ and is such that, for almost all t, the equation in u
$$\frac{\partial H}{\partial u}(\tilde{\psi}(t), x(t), u, t) = 0$$
has a unique solution in the ε-neighborhood of the point $\tilde{u}(t)$. Therefore, this solution coincides with $\tilde{u}(t)$. In the case of an open $U \subset R^r$, we shall say that the optimal problem (1.3) is regular if its Hamiltonian
$$H(t, x, \psi, u) = \psi f(t, x, u)$$
satisfies the condition
$$\left|\det\left(\frac{\partial^2 H}{\partial u^i \partial u^j}(t, x, \psi, u)\right)\right| \geq \alpha > 0 \qquad (1.17)$$
for an arbitrary ψ with $|\psi| = 1$, and for any t, x, and u that lie in a bounded part of the space R^{1+n+r} (α can depend on the bounded part of R^{1+n+r}).

Clearly, it follows from what we have said that, with initial conditions for x and ψ given, extremals of the regular optimal problem (1.3) are determined uniquely as the solutions of the canonical system
$$\dot{x} = \frac{\partial H(t, x, \psi, u)}{\partial \psi},$$
$$\dot{\psi} = -\frac{\partial H(t, x, \psi, u)}{\partial x}, \qquad (1.18)$$

from which u is eliminated with the aid of the equation

$$\frac{\partial H}{\partial u}(t, x, \psi, u) = 0. \tag{1.19}$$

This equation determines u in a unique way in a neighborhood of any of its solutions as a function

$$u = u(t, x, \psi),$$

which is differentiable with respect to t, x, ψ.

The elimination of u from the canonical system (1.18) can be considered as a transformation of the "initial variables" x, u of the problem (1.3) to the new, *canonically adjoint* variables x and ψ which satisfy the canonical system (1.18). In this connection, the vector ψ, canonically adjoint to the vector x, is determined implicitly from the formula (1.19) to within a constant scalar factor. This transformation from x, u to $x, \lambda\psi$ can be accomplished successfully for the regular problems [when the condition (1.17) is satisfied]. It is called the Legendre transformation generated by the function $f(t, x, u)$.

From this viewpoint, the Pontryagin maximum condition (1.9) is a Legendre transformation generalized to the case where the set U of admissible values of the control is arbitrary, and the function $f(t, x, u)$ is not necessarily differentiable with respect to u.

We have already described the geometrical sense of this generalized Legendre transformation in Section 1.4, when we gave the geometrical interpretation of the maximum condition. Namely, in order to find the values of u that correspond to given t, x, and ψ, one moves the hyperplane Γ^{n-1} orthogonal to ψ in R^n in the direction of the vector ψ until it intersects the set

$$P(t, x) = \{f(t, x, u): u \in U\}$$

for the last time at the (unique by assumption) point $\hat{f} = f(t, x, \hat{u})$. The value $u = \hat{u}$ is the required value [we assume that the equation $\hat{f} = f(t, x, u)$ is uniquely solvable in $u \in U$].

Conversely, in order to find ψ with $|\psi| = 1$ that corresponds to t, x, and u, one draws the support hyperplane to the convex hull of the set $P(t, x)$ through the point $f(t, x, u)$ (by assumption, such a hyperplane exists and is unique), and one takes the vector ψ orthogonal to this hyperplane and directed from the set $P(t, x)$.

The method of finding optimal solutions with the aid of the maximum principle described in this chapter is called the *canonical formalism*. Here is a brief summary of this method:

We form the Hamiltonian $H = \psi f(t, x, u)$ of the problem (1.3). With the aid of this function, we write the canonical system (1.7). The solutions of this system are obtained by the elimination of u with the aid of the Pontryagin maximum condition (1.9). The solutions which satisfy the inequality (1.10) and the boundary conditions (1.11) form the set of extremals. The required optimal solutions are found among these extremals.

In many cases of importance, there exists a unique extremal of the problem (1.3), which therefore is the optimal solution—if such a solution exists at all.

The problem of the existence of optimal solutions will be considered in Chapter 8.

1.7. Concluding Remarks

The maximum principle is a *first-order necessary condition*. It is obtained as a result of an investigation of the cone of (first-order) variations of an optimal trajectory of a certain control problem that uniquely corresponds to the initial optimal problem. This control problem is called a *convex control problem*.

Convex control problems play an important role in control theory. In order to define them precisely, it is convenient to introduce the notion of *generalized control*. Chapters 2 and 3 are devoted to the definition of generalized controls and convex control problems, and to a study of their basic properties.

The main advantages of generalized controls over ordinary controls become apparent when one *varies* the trajectories of controlled equations and when one constructs the cone of variations of these trajectories. The corresponding constructions are described in Chapter 6. In Chapter 7, the proof of the maximum principle is given on the basis of the results of Chapter 6.

Another significant advantage of generalized controls over ordinary controls is their weak compactness. This property is studied in Chapter 8, and is used to prove certain basic facts concerning the existence of optimal solutions.

Finally, Chapters 4 and 5 are auxiliary in character. In these chapters, we formulate and prove those general theorems on the existence of solutions of ordinary differential equations and on the dependence of these solutions on the initial data and right-hand sides that we use in this course.

2

Generalized Controls

In this chapter, we expand the class of admissible controls by the introduction of generalized controls. Advantages of this expansion will become clear when we study the method of variations in control problems and the proof of the maximum principle based on this method (Chapters 6 and 7), and also when we prove the existence of an optimal solution (Chapter 8).

Here, we shall prove the basic properties of generalized controls, paying particular attention to the notion of weak convergence of generalized controls, a notion which is fundamental for us.

2.1. Generalized Controls and a Convex Control Problem

We begin with two definitions: Let μ_t, $t \in R$ be a family of Radon measures on R^r that depend on the parameter $t \in R$. If $g(t, u)$ is a continuous (scalar- or vector-valued) function of its arguments $t \in R$ and $u \in R^r$ with a compact support in u for every fixed $t \in R$ (the support can depend on t), then, integrating $g(t, u)$ with respect to μ_t, we obtain the following function of t:

$$h(t) = \int_{R^r} g(t, u) d\mu_t(u) = \int_{R^r} g(t, u) d\mu_t, \qquad t \in R.$$

If the function $h(t)$ is Lebesgue measurable for an arbitrary $g(t, u)$ of this type, then we say that the family μ_t, $t \in R$, is *weakly measurable* (with respect to t).

If there exists a compact set $K \subset R^r$ that does not depend on $t \in R$ and is such that the measures μ_t are concentrated on K for almost all $t \in R$ (in the sense of the Lebesgue measure on R), then the family μ_t, $t \in R$, is said to be *finite*.

We shall denote the result of the integration of a continuous function $g(t, u)$ with respect to a measure μ_t by

$$\langle \mu_t, g(t, u) \rangle = \int_{R^r} g(t, u) d\mu_t.$$

An admissible control $u(t) \in \Omega_U$ can be considered as a family of Dirac measures (a Dirac measure is a unit, positive measure concentrated at a point) on R^r that depend on time $t \in R$. Indeed, to the value $u(t)$ of the control at the time t, there corresponds the unit, positive measure $\delta_{u(t)}$ which is concentrated at the point $u(t) \in U$ and acts on an arbitrary continuous function $g(t, u)$ in accordance with the formula

$$\langle \delta_{u(t)}, g(t, u) \rangle = \int_{R^r} g(t, u) d\delta_{u(t)} = g(t, u(t)).$$

Since we consider only bounded controls which take on values in U, the set

$$N = \{u(t): t \in R\} \subset U \subset R^r$$

is bounded and, therefore, all measures of the family $\delta_{u(t)}$, $t \in R$, are concentrated on N. In other words, the family of measures $\delta_{u(t)}$ is finite.

Since the functions $u(t) \in \Omega_U$ are measurable, the function

$$h(t) = \langle \delta_{u(t)}, g(t, u) \rangle = \int_{R^r} g(t, u) d\delta_{u(t)} = g(t, u(t))$$

is measurable for any continuous function $g(t, u)$, i.e., the family of measures $\delta_{u(t)}$ is weakly measurable. Thus, we see that, for any admissible control $u(t) \in \Omega_U$, the corresponding family of Dirac measures $\delta_{u(t)}$ is finite and weakly measurable.

Conversely, assume that we have an arbitrary, weakly measurable finite family of Dirac measures $\delta_{v(t)}$, $t \in R$, where the measure $\delta_{v(t)}$ is concentrated at the point $v(t) \in U$ at the time t. Then it follows directly from the definitions given that the function $v(t)$, $t \in R$, is essentially bounded. Setting $g(t, u) = u$, we obtain the measurable function

$$\langle \delta_{v(t)}, u \rangle = v(t) \in U.$$

Thus, we have established a natural correspondence between admissible

Generalized Controls

controls $u(t) \in \Omega_U$ and weakly measurable and finite families of Dirac measures $\delta_{u(t)}$, $t \in R$, concentrated on the set $U \subset R^r$.

We now give the following basic definition:

Any weakly measurable and finite family of *probability measures*, i.e., unit, positive, Radon measures μ_t with $t \in R$ that are concentrated on the set $U \subset R^r$, is said to be a *generalized control*.

We denote the set of all generalized controls by $\mathfrak{M}_U$ and call it the *class of generalized controls*. Subsequently, μ_t with $t \in R$ will always denote a generalized control.

The reason for taking a probability measure, and not an arbitrary Radon measure in the definition of a generalized control, is that only families of probability measures have the property that makes them useful in control problems and that is expressed in the approximation lemma. (See Chapter 3.)

The function resulting from the integration of $g(t, u)$ with respect to a measure μ_t, i.e.,

$$\langle \mu_t, g(t, u) \rangle = \int_{R^r} g(t, u) d\mu_t, \quad t \in R,$$

is called the *substitution of the generalized control μ_t for the parameter u in $g(t, u)$*. This agrees with the usual terminology, since the substitution of the function $u(t)$ into $g(t, u)$ leads to the same result as the substitution of the corresponding family of Dirac measures $\delta_{u(t)}$, $t \in R$. Henceforth we identify $u(t) \in \Omega_U$ with $\delta_{u(t)}$, $t \in R$, i.e., we assume that $\Omega_U \subset \mathfrak{M}_U$.

We now return to the controlled equation

$$\dot{x} = f(t, x, u).$$

Substituting a generalized control μ_t for u on the right-hand side of this equation, we obtain the differential equation

$$\dot{x} = \langle \mu_t, f(t, x, u) \rangle = \int_{R^r} f(t, x, u) d\mu_t = F(t, x),$$

which is analogous to equation (1.2). If the initial condition $x(t_1) = x_1$ is given, then the equation obtained is equivalent to the integral equation

$$x(t) = x_1 + \int_{t_1}^{t} \langle \mu_\theta, f(\theta, x(\theta), u) \rangle d\theta = x_1 + \int_{t_1}^{t} F(\theta, x(\theta)) d\theta,$$

which has a uniquely determined solution defined on a neighborhood of the point t_1 (see Chapter 4). The fact that now we substitute not only ordinary controls but also arbitrary generalized controls will be indicated in the following form:

$$\dot{x} = \langle \mu_t, f(t, x, u) \rangle, \qquad \mu_t \in \mathfrak{M}_U.$$

As before, the following boundary value problem can be formulated: Find a generalized control μ_t such that the solution of the equation

$$\dot{x} = \langle \mu_t, f(t, x, u) \rangle$$

satisfies the boundary conditions

$$x(t_1) = x_1, \qquad x(t_2) = x_2.$$

If the condition

$$t_2 - t_1 \to \min$$

is added to this boundary value problem, then we obtain an optimal problem in the class of generalized controls.

The control problem

$$\dot{x} = \langle \mu_t, f(t, x, u) \rangle, \qquad \mu_t \in \mathfrak{M}_U, \tag{2.1}$$

is said to be the *convex control problem* which corresponds to, or is the *convexification* of, the problem

$$\dot{x} = f(t, x, u), \qquad u(t) \in \Omega_U. \tag{2.2}$$

Accordingly, the optimal problem

$$\begin{aligned} &\dot{x} = \langle \mu_t, f(t, x, u) \rangle, \qquad \mu_t \in \mathfrak{M}_U, \\ &t = t_1, \qquad x(t_1) = x_1, \qquad x(t_2) = x_2, \\ &t_2 - t_1 \to \min \end{aligned} \tag{2.3}$$

will be called the *convex optimal problem* which corresponds to the optimal problem (1.3).

Since Ω_U is contained in $\mathfrak{M}_U$, the class $\mathfrak{M}_U$ together with equation (2.1) determine a controlled object with wider control possibilities than the controlled object defined with the aid of the class Ω_U of usual controls. Therefore, generally speaking, the initial optimal problem (1.3) can have no solutions even though the convex optimal problem (2.3) is solvable. We shall study this question in the chapter devoted to the existence of optimal solutions (Chapter 8). Here, we shall indicate the geometrical sense of the transfer to the convex control problem, which justifies its name.

Generalized Controls 25

First of all, it is clear that the set of all generalized controls $\mathfrak{M}_U$ is convex. Indeed, if μ_t and v_t are two generalized controls, then for any constants α and β the family of measures $\alpha\mu_t + \beta v_t$ is weakly measurable, finite, and concentrated on U for any t, because such are μ_t and v_t. If, moreover, the constants α and β satisfy the conditions $\alpha \geq 0$, $\beta \geq 0$, and $\alpha + \beta = 1$, then the measure $\alpha\mu_t + \beta v_t$ is a probability measure for every t. Furthermore, it is clear that the set of the right-hand sides of the convex control problem (2.1) is also convex. We shall explain this assertion in more detail.

The set of the right-hand sides of the initial control problem (2.2) can be written in the form

$$\{F(t, x): F(t, x) = f(t, x, u(t)), u(t) \in \Omega_U\}.$$

This is a complicated family of functions that depends on the set of admissible values U, and it certainly does not have to be convex in the linear space of all vector-valued n-dimensional functions $F(t, x)$ which are continuously differentiable with respect to x and measurable in t. The set of the right-hand sides of the corresponding convex problem has the form

$$\{F(t, x): F(t, x) = \langle \mu_t, f(t, x, u) \rangle, \mu_t \in \mathfrak{M}_U\} \qquad (2.4)$$

and is always convex, independently of the form of the set U. Indeed, the function

$$\alpha F_1(t, x) + \beta F_2(t, x) = \langle \alpha\mu_t + \beta v_t, f(t, x, u) \rangle, \qquad \alpha, \beta \geq 0, \quad \alpha + \beta = 1,$$

belongs to this set together with the functions

$$F_1(t, x) = \langle \mu_t, f(t, x, u) \rangle, \qquad F_2(t, x) = \langle v_t, f(t, x, u) \rangle$$

by the convexity of $\mathfrak{M}_U$. In particular, it follows that the set of all possible phase velocities of the convex problem (2.1) with fixed t and x is also convex in R^n. We denote this set by $P_{\mathfrak{M}}(t, x)$. It is easy to see that

$$P_{\mathfrak{M}}(t, x) = \{p: p = \langle \mu, f(t, x, u) \rangle\},$$

where μ is an arbitrary probability measure concentrated on a finite part of U, because the family $\mu_t \equiv \mu$ with $t \in R$ is a generalized control.

Assertion 2.1. For arbitrary t and x, the set of all possible phase velocities $P_{\mathfrak{M}}(t, x)$ of the convex control problem (2.1) coincides with the convex hull of the set $P(t, x)$ of all possible phase velocities (for the values of t and x under consideration) of the initial control problem (2.2),

$$P_{\mathfrak{M}}(t, x) = \operatorname{conv} P(t, x) = \operatorname{conv} \{p: p = f(t, x, u), u \in U\},$$

where conv $\{\cdot\}$ denotes the convex hull of the set in braces.

Proof. Since the set $P_{\mathfrak{M}}(t, x)$ is convex and

$$P_{\mathfrak{M}}(t, x) \supset P(t, x),$$

it remains only to prove the inclusion

$$P_{\mathfrak{M}}(t, x) \subset \text{conv } P(t, x).$$

Let there be a point $\hat{p} \notin \text{conv } P(t, x)$ and let there exist, nevertheless, a probability measure $\hat{\mu}$ concentrated on U and such that

$$\langle \hat{\mu}, f(t, x, u) \rangle = \hat{p}.$$

We draw in R^n an $(n-1)$-dimensional hyperplane Γ^{n-1} through the point $\hat{p}$ in such a way that the convex set conv $P(t, x)$ lies on one side of it. We denote the characteristic function of the inverse image of the hyperplane Γ^{n-1} under the mapping

$$f(t, x, \cdot): R^r \to R^n \tag{2.5}$$

by $\chi_1(u)$, and the characteristic function of the complement by $\chi_2(u)$:

$$\chi_1(u) = \begin{cases} 1 & \text{for } f(t, x, u) \in \Gamma^{n-1}, \\ 0 & \text{for } f(t, x, u) \notin \Gamma^{n-1}, \end{cases}$$

$$\chi_2(u) = 1 - \chi_1(u).$$

Since the measure $\hat{\mu}$ is a probability measure, we have

$$\hat{p} = \langle \hat{\mu}, \hat{p} \rangle,$$

where, by $\hat{p}$ on the right, we mean the function that assumes the constant value $\hat{p}$ on U. Therefore,

$$\langle \hat{\mu}, f(t, x, u) - \hat{p} \rangle = \langle \hat{\mu}, \chi_1(u)(f(t, x, u) - \hat{p}) \rangle + \langle \hat{\mu}, \chi_2(u)(f(t, x, u) - \hat{p}) \rangle = 0.$$

We denote by ξ the vector orthogonal to Γ^{n-1} and directed toward the set conv $P(t, x)$. Using the scalar product with the vector ξ in the last equality, we obtain

$$\langle \hat{\mu}, \chi_1(u)\xi \cdot (f(t, x, u) - \hat{p}) \rangle + \langle \hat{\mu}, \chi_2(u)\xi \cdot (f(t, x, u) - \hat{p}) \rangle = 0.$$

The first term on the left-hand side is zero, because for those u for which $\chi_1(u) \neq 0$ the scalar product $\xi \cdot (f(t, x, u) - \hat{p}) = 0$ and, therefore,

$$\langle \hat{\mu}, \chi_2(u)\xi \cdot (f(t, x, u) - \hat{p}) \rangle = 0.$$

Generalized Controls

Since the measure $\hat{\mu}$ is positive, and since the scalar-valued function

$$\chi_2(u)\xi \cdot (f(t, x, u) - \hat{p}) > 0$$

for $f(t, x, u) \notin \Gamma^{n-1}$ (because the set conv $P(t, x)$ lies on the one side of Γ^{n-1}), we conclude that the measure $\hat{\mu}$ is concentrated on the inverse image of Γ^{n-1} under the mapping (2.5).

The same arguments, only with R^n replaced by Γ^{n-1} and the set conv $P(t, x)$ replaced by $\Gamma^{n-1} \cap \text{conv } P(t, x)$, will convince us that the measure $\hat{\mu}$ is concentrated on the inverse image of an $(n-2)$-dimensional hyperplane $\Gamma^{n-2} \subset \Gamma^{n-1} \subset R^n$ under the mapping (2.5), where $\hat{p} \in \Gamma^{n-2}$. Reducing the dimensionality in this way, we find that the measure $\hat{\mu}$ is concentrated on the inverse image of the point $\hat{p}$ under the mapping (2.5). However, this contradicts the assumptions that the positive measure $\hat{\mu}$ is concentrated on U and

$$\hat{p} \notin \text{conv } P(t, x),$$

since it follows from the last relation that the set U does not intersect the inverse image of $\hat{p}$ under the mapping (2.5).

The following important example of a generalized control is, in a sense, typical (see the approximation lemma in Chapter 3). For reasons which will become clear after the proof of the existence theorem (Chapter 8), this control is called the *sliding regime control*, or *chattering control* (see also the remark on terminology at the end of Chapter 3).

Let $u_1(t), \ldots, u_q(t)$ be an arbitrary number of admissible controls from Ω_U, and let $\mu_1(t), \ldots, \mu_q(t), t \in R$, be the same number of real-valued (Lebesgue) measurable functions that satisfy the conditions

$$\mu_i(t) \geqslant 0 \quad \text{and} \quad \sum_{i=1}^{q} \mu_i(t) = 1$$

for (almost) all t. These functions determine the generalized control

$$\mu_t = \sum_{i=1}^{q} \mu_i(t) \delta_{u_i(t)}, \quad t \in R, \tag{2.6}$$

which is called the *chattering control*. At every instant of time, the measure μ_t is concentrated at the points $u_1(t), \ldots, u_q(t)$ with the corresponding "weight coefficients" $\mu_1(t), \ldots, \mu_q(t)$. The control μ_t is, so to speak, "distributed" among the q points $u_1(t), \ldots, u_q(t)$ in the proportions $\mu_1(t), \ldots, \mu_q(t)$. For $q = 1$, we obtain an ordinary control

$$\mu_t = \delta_{u_1(t)} \in \Omega_U.$$

If we use a chattering control, then the controlled equation takes on the form

$$\dot{x} = \left\langle \sum_{i=1}^{q} \mu_i(t)\delta_{u_i(t)}, f(t, x, u) \right\rangle = \sum_{i=1}^{q} \mu_i(t) f(t, x, u_i(t)).$$

2.2. Weak Convergence of Generalized Controls

The notion of weak convergence is introduced in the linear space of all Radon measures on R^m with $m \geq 1$. We shall now remind the reader of this notion, and we shall make use of it in order to define the notion of weak convergence in the class of generalized controls.

A sequence of Radon measures v^i, $i = 1, 2, \ldots$, given on R^m is said to *converge weakly* to a measure v as $i \to \infty$ if, for any (scalar- or vector-valued) continuous function $g(z)$ on R^m with a compact support, the sequence

$$\langle v^{(i)}, g(z) \rangle = \int_{R^m} g(z) \, dv^{(i)}(z)$$

converges to $\langle v, g(z) \rangle$ as $i \to \infty$:

$$\langle v^{(i)}, g(z) \rangle \to \langle v, g(z) \rangle \qquad (i \to \infty).*$$

For a fixed t, an arbitrary generalized control μ_t, $t \in R$, is a probability measure on R^r. It can be viewed as a Radon measure v on the $(1+r)$-dimensional space R^{1+r} of points (t, u), if we define the action of the measure $v = \{\mu_t; t \in R\}$ on a continuous function $g(t, u)$ with a compact support in R^{1+r} by the formula

$$\langle v, g(t, u) \rangle = \int_R dt \int_{R^r} g(t, u) \, d\mu_t(u) = \int_R \langle \mu_t, g(t, u) \rangle \, dt,$$

---

*If we introduce a locally convex topology on the linear space of all Radon measures on R^m with the aid of the family of seminorms $\|v\|_g^* = |\langle v, g(z) \rangle|$, where $g(z)$ is a continuous function with a compact support in R^m, then the weak convergence is, obviously, the convergence in this topology. It can be shown that the topology thus defined does not have a countable basis of neighborhoods of the origin and, therefore, is nonmetrizable.

Generalized Controls

where the integration is carried out with respect to the Lebesgue measure, and the "scalar product" $\langle \mu_t, g(t, u) \rangle$ denotes, as indicated, the integration with respect to the measure μ_t for a fixed t.

We shall say that a sequence of generalized controls $\mu_t^{(i)}$ *converges weakly to a generalized control* μ_t as $i \to \infty$ if we have

$$\int_R \langle \mu_t^{(i)}, g(t, u) \rangle \, dt \to \int_R \langle \mu_t, g(t, u) \rangle \, dt \qquad (i \to \infty) \tag{2.7}$$

for an arbitrary continuous function $g(t, u)$ with compact support.

In connection with this definition, it is important to note the following: There is a generally accepted definition of weak convergence for a sequence of r-dimensional (Lebesgue) integrable functions $v_i(t)$, $t_1 \leq t \leq t_2$. The sequence $v_i(t)$ is said to converge weakly to $v(t)$, $t_1 \leq t \leq t_2$, as $i \to \infty$ if we have

$$\int_{t_1}^{t_2} \varphi(t) v_i(t) \, dt \to \int_{t_1}^{t_2} \varphi(t) v(t) \, dt \qquad (i \to \infty) \tag{2.8}$$

for any r-dimensional continuous row $\varphi(t)$, $t_1 \leq t \leq t_2$.

If we consider a sequence of ordinary controls,

$$\mu_t^{(i)} = \delta_{u_i(t)}, \qquad t \in R,$$

where we assume that the functions $u_i(t)$ with $t \in R$ are uniformly bounded in i, then we can investigate the convergence of $u_i(t)$ to a limit in the sense of definition (2.7)—meaning the weak convergence of generalized controls $\delta_{u_i(t)}$ to a generalized control. However, we can also investigate the convergence of the same sequence $u_i(t)$ (on an arbitrary interval $t_1 \leq t \leq t_2$) in the sense of definition (2.8). These are two completely different types of convergence, which are clearly shown by the following typical example:

For simplicity, we consider a scalar case ($r = 1$), and assume that we are given a sequence of partitions of the entire t-axis into intervals of the same length, where the length of intervals of the ith partition tends to zero as $i \to \infty$. A function $u_i(t)$ with $t \in R$ is defined with the aid of the ith partition as follows: This function is equal to either 1 or -1 on each of the partition intervals, and it jumps successively from one value to the other as the argument t changes from one interval to the next. For large i, $u_i(t)$ is a function that "rapidly oscillates" between 1 and -1, remaining at each value for the same length of time. It is then easy to see that, for any continuous function $\varphi(t)$, $t_1 \leqslant t \leqslant t_2$,

$$\int_{t_1}^{t_2} \varphi(t) u_i(t)\, dt \to 0 \qquad (i \to \infty),{}^*$$

i.e., $u_i(t) \to 0$ $(i \to \infty)$ in the sense of (2.8).

However, the sequence $\delta_{u_i(t)}$ does not tend to zero at all in the sense of the convergence of generalized controls (2.7). As we shall see later, we have

$$\int_{t_1}^{t_2} \langle \delta_{u_i(t)}, g(t, u) \rangle\, dt = \int_{t_1}^{t_2} g(t, u_i(t))\, dt$$

$$\to \frac{1}{2} \int_{t_1}^{t_2} g(t, 1)\, dt + \frac{1}{2} \int_{t_1}^{t_2} g(t, -1)\, dt \qquad (i \to \infty) \quad (2.9)$$

for any continuous function $g(t, u)$. Therefore, if $g(t, u)$ has a compact support in R^{1+r}, then

$$\int_R \langle \delta_{u_i(t)}, g(t, u) \rangle\, dt \to \frac{1}{2} \int_R g(t, 1)\, dt + \frac{1}{2} \int_R g(t, -1)\, dt$$

$$= \int_R \langle \mu_t, g(t, u) \rangle\, dt \qquad (i \to \infty), \qquad (2.10)$$

where

$$\mu_t = \tfrac{1}{2}\delta_1 + \tfrac{1}{2}\delta_{-1}.$$

We shall not present here the easy proof of formula (2.9), since it follows directly from a much more general relation which will be obtained in proving the approximation lemma in the next chapter.

Relation (2.10) states that the sequence of generalized controls $\delta_{u_i(t)}$

---

*This relation can be easily obtained if the function $\varphi(t)$ is continuously differentiable:

$$\int_{t_1}^{t_2} \varphi(t) u_i(t)\, dt = \left\{ \varphi(t_2) \int_{t_1}^{t_2} u_i(t)\, dt - \int_{t_1}^{t_2} \varphi'(t) \left(\int_{t_1}^{t} u_i(\theta)\, d\theta \right) dt \right\} \to 0 \qquad (i \to \infty).$$

Approximating a continuous function by a continuously differentiable function with arbitrary accuracy (e.g., by the Weierstrass theorem on the approximation with polynomials), we obtain the desired result.

Generalized Controls

converges weakly [in the sense of (2.7)] to a generalized control,

$$\delta_{u_i(t)} \to \tfrac{1}{2}\delta_1 + \tfrac{1}{2}\delta_{-1} \qquad (i \to \infty).$$

The larger the index i, the larger the frequency with which the control $u_i(t)$ "chatters" between the values 1 and -1, remaining at each value for the same amount of time, and the "closer" this control is [in the sense of the weak convergence (2.7)] to the chattering control $\tfrac{1}{2}\delta_1 + \tfrac{1}{2}\delta_{-1}$, which is distributed between the two points 1 and -1 with equal weight coefficients of $\tfrac{1}{2}$. This example motivates the terminology that we introduced for controls of the form (2.6) (see also the end of Chapter 3). The problem of weak convergence of generalized controls will be studied in more detail in Chapter 8, which is devoted to the existence of optimal solutions.

Making use of the notion of strong convergence of general Radon measures, we shall now define the notion of strong convergence of generalized controls in a way similar to that used for weak convergence. We denote by $C^0(R^m)$ the normed space of all scalar-valued functions $g(z)$ continuous on R^m, with compact supports, and with the norm given by

$$\|g(z)\| = \sup_{z \in R^m} |g(z)|.$$

Convergence in this norm corresponds to uniform convergence.

We shall denote the *norm* or *total variation* of a Radon measure v on R^m by $\|v\|$,

$$\|v\| = \sup\{\langle v, g(z) \rangle : \|g(z)\| \leq 1, g(z) \in C^0(R^m)\}.$$

Obviously, a linear combination of probability measures or a measure with a compact support has a finite norm.

It is not difficult to see that, if v_t, $t \in R$, is a weakly measurable family of Radon measures on R^m, then the function

$$h(t) = \|v_t\|, \qquad t \in R,$$

is Lebesgue measurable. Indeed, we choose in $C^0(R^m)$ a countable sequence of functions $g_1(z), g_2(z), \ldots$ that satisfies the following conditions: (1) $\|g_i(z)\| \leq 1$ for all $i = 1, 2, \ldots$, and (2) for every $g(z) \in C^0(R^m)$ with the norm $\|g(z)\| \leq 1$, one can choose a subsequence $g_{i_k}(z)$ such that

$$\|g_{i_k}(z) - g(z)\| \to 0 \qquad (k \to \infty),$$

where all the functions $g_{i_k}(z)$ have supports that lie in a bounded part of the space R^m. The existence of such a sequence will be proved in Chapter 3 (Assertion 3.3).

Taking into account the fact that all the functions $g(z)$ and $g_{i_k}(z)$ have supports that lie in a bounded part of R^m, we obtain

$$\langle v_t, g_{i_k}(z) - g(z) \rangle \to 0 \quad (k \to \infty) \quad \forall t \in R.$$

Hence

$$\langle v_t, g(z) \rangle \leq \sup_i \langle v_t, g_i(z) \rangle \quad \forall t \in R.$$

However, since $g(z)$ is an arbitrary function from $C^0(R^m)$ with the norm $\|g(z)\| \leq 1$, and since all the norms $\|g_i(z)\| \leq 1$, we have

$$\|v_t\| \leq \sup_i \langle v_t, g_i(z) \rangle \leq \|v_t\|, \quad t \in R,$$

i.e.,

$$h(t) = \|v_t\| = \sup_i \langle v_t, g_i(z) \rangle \quad \forall t \in R.$$

The function $\|v_t\|$, $t \in R$, is the upper bound of the countable sequence of functions

$$h_i(t) = \langle v_t, g_i(z) \rangle, \quad t \in R.$$

By the assumed weak measurability of the family v_t, the functions $h_i(t)$ are Lebesgue measurable and therefore so is $\|v_t\|$.

We shall say that a sequence of Radon measures $v^{(i)}$, $i = 1, 2, \ldots$, given on R^m converges strongly to a measure v as $i \to \infty$ if

$$\|v^{(i)} - v\| \to 0 \quad (i \to \infty).$$

Furthermore, we shall say that a sequence of generalized controls $\mu_t^{(i)}$, $t \in R$, converges strongly to a generalized control μ_t, $t \in R$, if

$$\int_R \|\mu_t^{(i)} - \mu_t\| \, dt \to 0 \quad (i \to \infty),$$

where the norm under the integral denotes the norm of the Radon measure within the double bars on R^r for a given t, and the integral on R is taken, as in (2.7), with respect to Lebesgue measure.

Obviously, as in the case of general Radon measures, the strong convergence of a sequence of generalized controls $\mu_t^{(i)}$ to a generalized control μ_t implies the weak convergence, but the converse is not true.

Here is a simple example of a sequence of generalized controls which converges weakly, but not strongly: Let a sequence of points u_i in R^r tend to a point $\hat{u}$, where $u_i \neq \hat{u}$. We define

$$\mu_t^{(i)} = \delta_{u_i} \quad \text{and} \quad \mu_t = \delta_{\hat{u}} \quad \forall t \in R.$$

Generalized Controls

We have

$$\langle \delta_{u_i}, g(t, u)\rangle = g(t, u_i) \to g(t, \hat{u}) = \langle \delta_{\hat{u}}, g(t, u)\rangle.$$

Hence,

$$\int_R \langle \delta_{u_i}, g(t, u)\rangle \, dt \to \int_R \langle \delta_{\hat{u}}, g(t, u)\rangle \, dt \quad \text{as } i \to \infty,$$

i.e., $\delta_{u_i} \to \delta_{\hat{u}}$ $(i \to \infty)$ weakly, but not strongly, since $u_i \neq \hat{u}$ implies $\|\delta_{u_i} - \delta_{\hat{u}}\| = 2$.

In conclusion, we shall prove the following assertion, which we shall need in Chapter 6·

Assertion 2.2. Let there be given a sequence of generalized controls $\mu_t^{(i)}(\sigma)$, $i = 1, 2, \ldots$, that depend on a parameter $\sigma \in \Sigma$. We assume that the sequence converges weakly to a generalized control $\mu_t(\sigma)$, $\sigma \in \Sigma$, as $i \to \infty$ uniformly with respect to $\sigma \in \Sigma$, and that all the measures

$$\mu_t(\sigma) \quad \text{and} \quad \mu_t^{(i)}(\sigma), \quad t \in R, \sigma \in \Sigma, i = 1, 2, \ldots,$$

are concentrated on a single bounded set

$$N \subset U \subset R^r.$$

Then we have, for any continuous function $F(t, u)$, $(t, u) \in R \times R^r$, and for any numbers t' and t'',

$$\int_{t'}^{t''} \langle \mu_t^{(i)}(\sigma) - \mu_t(\sigma), F(t, u)\rangle \, dt \to 0 \quad (i \to \infty), \tag{2.11}$$

uniformly with respect to $\sigma \in \Sigma$.

Proof. We denote by S a closed ball in R^r with center at the origin and containing the set N. Let $\beta(u)$ be a scalar-valued function, continuous on R^r with a compact support, taking on values between 0 and 1, and identically equal to 1 on S. Such a function can be constructed as follows: We take a ball S' concentric with S and of a larger radius, and we set

$$\beta(u) = \begin{cases} 1 & \forall u \in S, \\ 0 & \forall u \notin S'. \end{cases}$$

In the layer between S and S', the function $\beta(u)$ decreases linearly from 1 to 0 along radii.

Similarly, let $\alpha(t)$ with $t \in R$ be a scalar-valued continuous function which is 1 on the interval $[t', t''](t' \leq t'')$, 0 outside of some interval $[t' - \eta, t'' + \eta]$, $0 < \eta \leq 1$, and linear on the intervals $[t' - \eta, t']$ and $[t'', t'' + \eta]$.

Since the measures $\mu_t^{(i)}(\sigma)$ and $\mu_t(\sigma)$ are concentrated on N, and since the function $\beta(u) = 1$ for all $u \in N$, we have

$$h^{(i)}(t;\sigma) = \langle \mu_t^{(i)}(\sigma) - \mu_t(\sigma), \alpha(t)\beta(u)F(t,u) \rangle = \alpha(t) \int_N \beta(u) F(t,u) \, d(\mu_t^{(i)}(\sigma) - \mu_t(\sigma))$$

$$= \alpha(t) \int_N F(t,u) \, d(\mu_t^{(i)}(\sigma) - \mu_t(\sigma)) = \langle \mu_t^{(i)}(\sigma) - \mu_t(\sigma), \alpha(t) F(t,u) \rangle.$$

The function $\alpha(t)\beta(u)F(t, u)$ has compact support. Therefore, we have, in accordance with the assumption,

$$\int_R \langle \mu_t^{(i)}(\sigma) - \mu_t(\sigma), \alpha(t)\beta(u)F(t,u) \rangle \, dt = \int_R \langle \mu_t^{(i)}(\sigma) - \mu_t(\sigma), \alpha(t) F(t,u) \rangle \, dt$$

$$= \int_R h^{(i)}(t;\sigma) \, dt \to 0 \quad (i \to \infty), \quad (2.12)$$

uniformly with respect to $\sigma \in \Sigma$. Moreover,

$$\int_{t'}^{t''} h^{(i)}(t;\sigma) \, dt = \int_{t'}^{t''} \langle \mu_t^{(i)}(\sigma) - \mu_t(\sigma), F(t,u) \rangle \, dt \quad \forall \sigma \in \Sigma \text{ and } \forall i = 1, 2, \ldots.$$

Since $\|\mu_t^{(i)}(\sigma) - \mu_t(\sigma)\| \leq 2$ and $0 \leq \alpha(t) \leq 1$, the following estimate holds:

$$|h^{(i)}(t;\sigma)| \leq \int_N |F(t,u)| \, d|\mu_t^{(i)}(\sigma) - \mu_t(\sigma)|$$

$$\leq \max_{(t,u) \in [t'-1, t''+1] \times S} |F(t,u)| \int_N d|\mu_t^{(i)}(\sigma) - \mu_t(\sigma)|$$

$$\leq 2 \max_{(t,u) \in [t'-1, t''+1] \times S} |F(t,u)| = C.$$

Thus,

$$\int_R \langle \mu_t^{(i)}(\sigma) - \mu_t(\sigma), \alpha(t)F(t, u)\rangle\, dt = \int_R h^{(i)}(t;\sigma)\, dt$$

$$= \int_{t'-\eta}^{t'} h^{(i)}(t;\sigma)\, dt + \int_{t''}^{t''+\eta} h^{(i)}(t;\sigma)\, dt$$

$$+ \int_{t'}^{t''} \langle \mu_t^{(i)}(\sigma) - \mu_t(\sigma), F(t, u)\rangle\, dt.$$

Hence,

$$\left|\int_{t'}^{t''} \langle \mu_t^{(i)}(\sigma) - \mu_t(\sigma), F(t, u)\rangle\, dt\right| \leq \left|\int_R h^{(i)}(t;\sigma)\, dt\right| + 2C\eta.$$

This estimate concludes the proof of the assertion, since the constant C does not depend on η, since $\eta > 0$ can be chosen arbitrarily small, and since the integral

$$\int_R h^{(i)}(t;\sigma)\, dt$$

tends to zero as $i \to \infty$ [see (2.12)], uniformly with respect to $\sigma \in \Sigma$.

3

The Approximation Lemma

In this chapter, Σ denotes an arbitrary metric space.

Let us be given a mapping

$$\sigma \mapsto \mu_t(\sigma), \qquad \sigma \in \Sigma,$$

of the space Σ into the set of generalized controls $\mathfrak{M}_U$, i.e., we are given a family of generalized controls

$$\{\mu_t(\sigma): \sigma \in \Sigma\} \subset \mathfrak{M}_U.$$

We shall say that this family is *weakly continuous*, or that it *weakly continuously depends on the parameter* σ, if, for any continuous function $g(t, u)$ with a compact support, $\sigma_i \to \hat{\sigma}$ $(i \to \infty)$ implies

$$\int_R \langle \mu_t(\sigma_i) - \mu_t(\hat{\sigma}), g(t, u) \rangle \, dt \to 0 \qquad (i \to \infty).$$

Replacing this condition by the condition

$$\int_R \|\mu_t(\sigma_i) - \mu_t(\hat{\sigma})\| \, dt \to 0 \qquad (i \to \infty),$$

we obtain the notion of a *strongly continuous family* of generalized controls.

Obviously, any strongly continuous family of generalized controls is weakly continuous, but the converse is not true.

The approximation lemma asserts than an arbitrary strongly continuous family of generalized controls $\mu_t(\sigma), \sigma \in \Sigma$, can be approximated with arbitrary accuracy in the sense of weak convergence by some strongly continuous family of usual controls $\delta_{u(t;\sigma)}$, and that this approximation will be uniform with respect to $\sigma \in \Sigma$. Therefore, ordinary controls are not only dense everywhere (in the sense of weak convergence) in the class of all generalized controls $\mathfrak{M}_U$, but they can also be used as uniform approximations of strongly continuous families of generalized controls to within any accuracy.

The precise statement and the proof of the corresponding theorem will be given in Section 3.2. We shall now describe the technique of function "smoothing" and the construction of "partition of unity," which is necessary for the proof of the theorem and important for the entire mathematical analysis. The technique will be formulated in the form of Theorem 3.1, which will be presented in the degree of generality sufficient for our purposes.

3.1. Partition of Unity

We define a scalar-valued, nonnegative infinitely differentiable function $s(u)$ of an r-dimensional argument $u \in R^r$ by the formula

$$s(u) = \begin{cases} \gamma \exp\left(\dfrac{1}{|u|^2 - 1}\right), & |u| < 1, \\ 0, & |u| \geq 1, \end{cases}$$

where a positive constant γ is chosen so that the following equality holds:

$$\int_{R^r} s(u)\, du = 1,$$

i.e., we set

$$\frac{1}{\gamma} = \int_{|u|<1} \exp\left(\frac{1}{|u|^2 - 1}\right) du.$$

Obviously, the function $s(u)$ is infinitely differentiable, symmetric with respect to the origin,

$$s(u) = s(-u) \quad \forall u \in R^r,$$

attains its maximum value equal to $\gamma \exp(-1)$ at the origin, and vanishes outside of the ball $|u| \leq 1$, which is its support.

With the aid of $s(u)$, we define a family of infinitely differentiable func-

The Approximation Lemma

tions of two variables $u, v \in R^r$, which depends on a parameter $\varepsilon > 0$, by the formula

$$s_\varepsilon(u,v) = \frac{1}{\varepsilon^r} s\left(\frac{u-v}{\varepsilon}\right).$$

This family will be called the *smoothing family of functions*. The reason for this name will become apparent from Assertion 3.1.

For any fixed ε and u, the function $s_\varepsilon(u, v)$ of v has the center of symmetry at the point u, where it attains its maximum value $(1/\varepsilon^r)\gamma \exp(-1)$, has the ball $|u - v| \leq \varepsilon$ as its support, and satisfies the relation

$$\int_{R^r} s_\varepsilon(u, v) \, dv = \int_{|u-v| \leq \varepsilon} s_\varepsilon(u, v) \, dv = 1 \qquad \forall u \in R \quad \text{and} \quad \forall \varepsilon > 0.$$

In order to prove the last equality, it is sufficient to perform the change of variables under the integral from v to $w = (u - v)/\varepsilon$, and note that the absolute value of the Jacobian is then equal to ε^r.

Let $g(u)$ be an arbitrary locally integrable function on R^r, i.e., a function which is integrable on an arbitrary bounded measurable (in the sense of r-dimensional Lebesgue measure) subset of R^r. We define a family of integral operators S_ε which depends on $\varepsilon > 0$ by the formula

$$g(u) \mapsto g_\varepsilon(u) = S_\varepsilon g(u) = \int_{R^r} s_\varepsilon(u, v) g(v) \, dv = \int_{|u-v| \leq \varepsilon} s_\varepsilon(u, v) g(v) \, dv.$$

This family will be called the *smoothing family of operators*, and the function $g_\varepsilon(u)$ will be called the ε-*smoothing* of the function $g(u)$.

Assertion 3.1. *For any locally integrable function $g(u)$, its ε-smoothing $g_\varepsilon(u)$ is infinitely differentiable for all $\varepsilon > 0$, and the partial derivatives are expressed by the formula*

$$\left(\frac{\partial}{\partial u^1}\right)^{k_1} \cdots \left(\frac{\partial}{\partial u^r}\right)^{k_r} g_\varepsilon(u) = \int_R \left(\frac{\partial}{\partial u^1}\right)^{k_1} \cdots \left(\frac{\partial}{\partial u^r}\right)^{k_r} s_\varepsilon(u, v) g(v) \, dv$$

$$\forall k_1 \geq 0, \ldots, k_r \geq 0.$$

Proof. Let an arbitrary scalar-valued function $\omega(u,v)$ be infinitely differentiable with respect to u for any fixed v, and let it vanish as a function of v outside of a fixed ball S when u remains in an arbitrary bounded set given in advance. We shall prove that, in this case, the function

$$h(u) = \int_{R^r} \omega(u, v)g(v)\, dv$$

is infinitely differentiable with respect to u, and that

$$\frac{\partial h}{\partial u^i} = \int_{R^r} \frac{\partial}{\partial u^i} \omega(u, v)g(v)\, dv, \qquad i=1,\ldots,r.$$

Hence the assertion follows directly, since partial derivatives of any order of the function $s_\varepsilon(u, v)$ have the enumerated properties.

We represent the difference $h(\hat{u}+\delta u) - h(\hat{u})$ in the form

$$h(\hat{u}+\delta u) - h(\hat{u}) = \int_{R^r} [\omega(\hat{u}+\delta u, v) - \omega(\hat{u}, v)]g(v)\, dv$$

$$= \int_{R^r} dv \int_0^1 \frac{\partial}{\partial u} \omega(\hat{u}+\theta\delta u, v)g(v)\, d\theta\, \delta u$$

$$= \int_S dv \int_0^1 \frac{\partial}{\partial u} \omega(\hat{u}+\theta\delta u, v)g(v)\, d\theta\, \delta u.$$

We have

$$h(\hat{u}+\delta u) - h(\hat{u}) - \int_{R^r} \frac{\partial}{\partial u} \omega(\hat{u}, v)g(v)\, dv\, \delta u$$

$$= \int_{R^r} dv \int_0^1 \left(\frac{\partial}{\partial u} \omega(\hat{u}+\theta\delta u, v) - \frac{\partial}{\partial u} \omega(\hat{u}, v) \right) g(v)\, d\theta\, \delta u$$

$$= \int_S \int_0^1 \left(\frac{\partial}{\partial u} \omega(\hat{u}+\theta\delta u, v) - \frac{\partial}{\partial u} \omega(\hat{u}, v) \right) g(v)\, dv\, d\theta\, \delta u.$$

Therefore,

$$\frac{1}{|\delta u|} \left| h(\hat{u}+\delta u) - h(\hat{u}) - \int_{R^r} \frac{\partial}{\partial u} \omega(\hat{u}, v)g(v)\, dv\delta u \right|$$

$$\leq \int_S \int_0^1 \left| \frac{\partial}{\partial u} \omega(\hat{u}+\theta\delta u, v) - \frac{\partial}{\partial u} \omega(\hat{u}, v) \right| \cdot |g(v)|\, dv\, d\theta.$$

The Approximation Lemma

The first factor of the integrand on the right-hand side tends to zero as $\delta u \to 0$ for any fixed θ and v, and is majorized by the constant

$$2 \max_{(u,v) \in V \times S} \left| \frac{\partial}{\partial u} \omega(u, v) \right|,$$

where V is a bounded neighborhood of the point $\hat{u}$. Therefore, we have on the basis of the Lebesgue theorem on passing to the limit under the integral that the left-hand side of the inequality tends to zero as $\delta u \to 0$, so that the assertion has been proved.

We shall need the following assertion in proving Theorem 3.1 on the partition of unity. It will also be used repeatedly in Chapters 4 and 5.

Assertion 3.2. Let there be given an arbitrary compact set $K \subset R^r$ and an arbitrary open set O which contains K. Then we can construct an infinitely differentiable function $\alpha(u)$ which has a compact support in O and satisfies the conditions

$$0 \leq \alpha(u) \leq 1 \quad \forall\, u \in R^r \quad \text{and} \quad \alpha(u) = 1 \quad \forall\, u \in K.$$

Proof. Let d be the distance from K to the closed set $R^r \backslash O$. Obviously, $d > 0$, since the set K is compact and does not intersect the closed set $R^r \backslash O$. We denote by V the $d/3$-neighborhood of the compact set K (if $d = \infty$, then V is an arbitrary bounded neighborhood of K), and let $\chi(u)$ be the characteristic function of the set V. As we know, the ε-smoothing of this function

$$\chi_\varepsilon(u) = \int_{R^r} s_\varepsilon(u, v) \chi(v)\, dv = \int_{|u-v| \leq \varepsilon} s_\varepsilon(u, v) \chi(v)\, dv,$$

is infinitely differentiable. Let us show that, for $\varepsilon \leq d/3$, the function $\chi_\varepsilon(u)$ can be taken for the required function. First, we have the following obvious estimate:

$$0 \leq \chi_\varepsilon(u) \leq \int_{R^r} s_\varepsilon(u, v)\, dv = 1.$$

Furthermore, it follows from $|u - v| \leq \varepsilon$ for $u \in K$ and $\varepsilon = d/3$ that $v \in V$, i.e., that $\chi(v) = 1$. Therefore, under these conditions,

$$\chi_\varepsilon(u) = \int_{|u-v| \leq \varepsilon} s_\varepsilon(u, v) \chi(v)\, dv = \int_{|u-v| \leq \varepsilon} s_\varepsilon(u, v)\, dv = 1.$$

Finally, if the distance from u to K is greater than $2d/3$, then $|u-v|\leq\varepsilon\leq d/3$ implies $v\notin V$, i.e., $\chi(v)=0$, and therefore

$$\chi_\varepsilon(u) = \int_{|u-v|\leq\varepsilon} s(u,v)\chi(v)\,dv = 0.$$

Thus, we can set

$$\alpha(u) = \chi_\varepsilon(u), \qquad 0 < \varepsilon \leq d/3.$$

It can be seen from the proof that we can take for V an arbitrarily small neighborhood of the compact set K lying in O. Therefore, an arbitrarily small positive number can be taken for ε. For this reason, we can conceive of a function $\alpha(u)$ which is identically 1 on K, sharply decreases to zero outside of K, and nevertheless remains infinitely differentiable, i.e., the infinitely differentiable function $\alpha(u)$ is "arbitrarily close" to the characteristic function of the compact set K.

Theorem 3.1. Let a compact set $K \subset R^r$ be covered by a finite number of open sets $O_1, \ldots, O_p$. Then there exist infinitely differentiable functions $\alpha_1(u), \ldots, \alpha_p(u)$ such that

(1) $0 \leq \alpha_i(u) \leq 1 \quad \forall u \in R^r, \quad i = 1, 2, \ldots, p;$

(2) the function $\alpha_i(u)$ has a compact support contained in O_i, $i = 1, \ldots, p$;

(3) $\sum_{i=1}^{p} \alpha_i(u) = 1 \quad \forall u \in K.$

The system of functions $\alpha_1(u), \ldots, \alpha_p(u)$ is said to be the *partition of unity* on the compact set K subordinate to the covering $O_1, \ldots, O_p$.

Proof. There exists a covering of the compact set K by bounded open sets $O'_1, \ldots, O'_p$ such that

$$\overline{O}'_i \subset O_i, \qquad i = 1, \ldots, p.$$

Indeed, we cover the compact set K by a finite number of open balls $S_1, \ldots, S_q$ in such a way that each of the closed balls $\overline{S}_j$ is contained in some O_i. We denote by O'_1 the union of all those S_j whose closures are contained in O_1. If there are no such balls, then we set O'_1 equal to an empty set. Obviously, $\overline{O}'_1 \subset O_1$. We denote by O'_2 the union of the balls S_j whose closures are con-

The Approximation Lemma

tained in O_2, etc. Obviously, the system of sets $O'_1, \ldots, O'_p$ is a required system.

According to Assertion 3.2, if O'_i is not empty, then we can construct a nonnegative infinitely differentiable function $\beta_i(u)$ equal to 1 on $\overline{O'_i}$ and with compact support in O_i. If O'_i is empty, then we set $\beta_i(u) \equiv 0$. We also have

$$\beta_1(u) + \cdots + \beta_p(u) > 0 \qquad \forall u \in \bigcup_{i=1}^{p} \overline{O'_i} \supset K,$$

because the corresponding function $\beta_i(u) = 1$ for $u \in \overline{O'_i}$. We also construct an infinitely differentiable function $\beta(u)$ equal to 1 on K and whose support is contained in the open set $\bigcup_{i=1}^{p} O'_i$. We define the functions $\alpha_i(u)$ by the formula

$$\alpha_i(u) = \begin{cases} \beta(u) \dfrac{\beta_i(u)}{\beta_1(u) + \cdots + \beta_p(u)}, & u \in \bigcup_{j=1}^{p} \overline{O'_j}, \\ 0, & u \notin \bigcup_{j=1}^{p} \overline{O'_j}. \end{cases}$$

It can be seen directly that these functions form a partition of unity on the compact set K subordinate to the covering $O_1, \ldots, O_p$.

We note that in the proof of the approximation lemma we shall need only the continuity of the functions $\alpha_i(u)$ which partition unity on K, but not the differentiability of these functions.

We shall now prove the following assertion, which we have already used in Chapter 1:

Assertion 3.3. Let $C^0(R^m)$ be the normed space of all scalar-valued continuous functions on R^m with compact supports, and with the norm of uniform convergence:

$$\|g(z)\| = \sup_{z \in R^m} |g(z)|.$$

Then there exists in $C^0(R^m)$ a sequence of functions $g_1(z), g_2(z), \ldots$ such that, for every $g(z) \in C^0(R^m)$, one can choose from this sequence a subsequence $g_{i_k}(z)$ that satisfies the following conditions:

(1) $\|g_{i_k}(z) - g(z)\| \to 0 \quad (k \to \infty)$;
(2) $\|g_{i_k}(z)\| \leq \|g(z)\| \quad \forall k = 1, 2, \ldots$;
(3) all functions $g_{i_k}(z)$, $k = 1, 2, \ldots$ have supports which lie in a bounded part of the space R^m.

Proof. We denote by

$$\{O_{l,k}^{(1)}, \ldots, O_{l,k}^{p_{l,k}}\}$$

the covering of the ball in R^m with radius l and center at the origin by open sets $O_{l,k}^{(i)}$, $i=1,\ldots,p_{l,k}$, whose diameters are less than or equal to $1/k$. We denote by

$$\{\alpha_{l,k}^{(i)}(z), i=1,\ldots,p_{l,k}\}$$

the partition of unity subordinate to the corresponding covering. We consider the countable set of functions

$$\sum_{i=1}^{p_{l,k}} r_i \alpha_{l,k}^{(i)}(z), \qquad l, k=1, 2, \ldots,$$

where r_i ranges over all rational numbers. We shall show that, numbering this set in an arbitrary way, we obtain a required sequence $g_1(z), g_2(z), \ldots$. Indeed, let $g(z)$ be an arbitrary function from $C^0(R^m)$, and let a number $\hat{l}$ be so large that the ball with radius $\hat{l}$ contains the support of the function $g(z)$. The function $g(z)$ is uniformly continuous, and the diameters of the sets $O_{\hat{l},k}^{(i)}$ tend to zero as $k \to \infty$ uniformly with respect to i. Therefore, for every $k=1, 2, \ldots$, we can choose rational numbers $r_k^{(i)}$, $i=1,\ldots,p_{\hat{l},k}$ such that the following relations hold:

$$|r_k^{(i)}| \leq \sup\{|g(z)|: z \in O_{\hat{l},k}^{(i)}\} \qquad \forall\, i=1, \ldots, p_{\hat{l},k},$$

$$\sup\{|r_k^{(i)} - g(z)|: z \in O_{\hat{l},k}^{(i)}\} \to 0 \qquad (k \to \infty),$$

where the convergence to zero is uniform with respect to i. In other words,

$$\max_i \sup\{|r_k^{(i)} - g(z)|: z \in O_{\hat{l},k}^{(i)}\} \to 0 \qquad (k \to \infty).$$

Let us prove that the sequence of functions

$$\sum_{i=1}^{p_{\hat{l},k}} r_k^{(i)} \alpha_{\hat{l},k}^{(i)}(z), \qquad k=1, 2, \ldots,$$

is a required subsequence for $g(z)$. Condition (1) follows from the relations

$$\left\| \sum_{i=1}^{p_{\hat{l},k}} r_k^{(i)} \alpha_{\hat{l},k}^{(i)}(z) - g(z) \right\| = \left\| \sum_i (r_k^{(i)} - g(z)) \alpha_{\hat{l},k}^{(i)}(z) \right\|$$

$$\leq \sup_{z \in R^m} \left(\sum_i |r_k^{(i)} - g(z)| \alpha_{\hat{l},k}^{(i)}(z) \right)$$

$$\leqslant \sup_{z\in R^m}\left(\sum_i(\sup_{w\in O_{i,k}^{(i)}}|r_k^{(i)}-g(w)|)\alpha_{i,k}^{(i)}(z)\right)$$

$$\leqslant \sup_{z\in R^m}\left(\left(\max_j \sup_{w\in O_{i,k}^{(j)}}|r_k^{(j)}-g(w)|\right)\sum_i \alpha_{i,k}^{(i)}(z)\right)$$

$$=\max_i \sup_{z\in O_{i,k}^{(i)}}|r_k^{(i)}-g(z)|\to 0 \qquad (k\to\infty).$$

Condition (2) follows from the relations

$$\left\|\sum_{i=1}^{p_{i,k}} r_k^{(i)}\alpha_{i,k}^{(i)}(z)\right\| = \sup_{z\in R^m}\left(\sum_i r_k^{(i)}\alpha_{i,k}^{(i)}(z)\right)$$

$$\leqslant \sup_{z\in R^m}\left(\sum_i |r_k^{(i)}|\alpha_{i,k}^{(i)}(z)\right)$$

$$\leqslant \sup_{z\in R^m}\left(\sum_i \left(\sup_{w\in O_{i,k}^{(i)}}|g(w)|\right)\alpha_{i,k}^{(i)}(z)\right)$$

$$\leqslant \|g(z)\|\sup_{z\in R^m}\sum_i \alpha_{i,k}^{(i)}(z) = \|g(z)\|.$$

Finally, condition (3) is satisfied in an obvious way.

3.2. The Approximation Lemma

Theorem 3.2. Let $\mu_t(\sigma)$, $\sigma\in\Sigma$, be a strongly continuous family of generalized controls, and let all the measures $\mu_t(\sigma)$, $t\in R$, $\sigma\in\Sigma$, be concentrated on a single bounded set $N\subset U\subset R^r$. Then we can construct a sequence of piecewise-constant controls

$$\delta_{u^{(i)}(t;\,\sigma)} \in \Omega_U, \qquad i=1,2,\ldots,$$

which depends on $\sigma\in\Sigma$ strongly continuously for every $i=1,2,\ldots$ and is such that (i) all the measures $\delta_{u^{(i)}(t;\,\sigma)}$ are concentrated on N [i.e., $u^{(i)}(t;\,\sigma)\in N$ for all $t\in R$ and $\sigma\in\Sigma$, $i=1,2,\ldots$], and (ii) the sequence $\delta_{u^{(i)}(t;\,\sigma)}$ converges weakly to $\mu_t(\sigma)$ as $i\to\infty$, uniformly with respect to $\sigma\in\Sigma$. In other words, for any continuous function $g(t,u)$ on $R\times R^r$ with compact support, we have

$$\int_R \langle \mu_t(\sigma)-\delta_{u^{(i)}(t;\,\sigma)},\, g(t,u)\rangle\, dt \to 0 \qquad (i\to\infty),$$

where the convergence to zero is uniform with respect to $\sigma\in\Sigma$.

Proof. Let

$$\{O_j^{(i)}, j=1, \ldots, p_i\}, \quad i=1, 2, \ldots,$$

be a sequence of coverings of the closure $\bar{N}$ of the set $N \subset R^r$ by bounded open sets that satisfies the following condition: The maximal diameter of the sets of the ith covering tends to zero as $i \to \infty$. We denote by

$$\{\alpha_j^{(i)}(u), j=1, \ldots, p_i\}$$

the partition of unity on the compact set $\bar{N}$ subordinate to the covering

$$O_1^{(i)}, \ldots, O_{p_i}^{(i)}.$$

For any $\sigma \in \Sigma$, the functions

$$\lambda_j^{(i)}(t; \sigma) = \langle \mu_t(\sigma), \alpha_j^{(i)}(u) \rangle, \quad t \in R, \tag{3.1}$$

are measurable in t and satisfy the conditions

$$0 \leq \lambda_j^{(i)}(t; \sigma) \leq 1, \quad \sum_{j=1}^{p_i} \lambda_j^{(i)}(t; \sigma) \equiv 1 \quad \forall \sigma \in \Sigma, \quad i=1, 2, \ldots,$$

because for every $\sigma \in \Sigma$ we have

$$0 \leq \lambda_j^{(i)}(t; \sigma) = \int_N \alpha_j^{(i)}(u) \, d\mu_t(\sigma) \leq \int_N d\mu_t(\sigma) = 1,$$

$$\sum_{j=1}^{p_i} \lambda_j^{(i)}(t; \sigma) = \left\langle \mu_t(\sigma), \sum_{j=1}^{p_i} \alpha_j^{(i)}(u) \right\rangle = \int_N \sum_{j=1}^{p_i} \alpha_j^{(i)}(u) \, d\mu_t(\sigma) = \int_N d\mu_t(\sigma) = 1.$$

We shall show that, for $\sigma \to \hat{\sigma}$,

$$\sum_{j=1}^{p_i} \int_R |\lambda_j^{(i)}(t; \sigma) - \lambda_j^{(i)}(t; \hat{\sigma})| \, dt \to 0, \quad i=1, 2, \ldots. \tag{3.2}$$

We have

$$\sum_{j=1}^{p_i} \int_R |\lambda_j^{(i)}(t; \sigma) - \lambda_j^{(i)}(t; \hat{\sigma})| \, dt = \sum_{j=1}^{p_i} \int_R |\langle \mu_t(\sigma) - \mu_t(\hat{\sigma}), \alpha_j^{(i)}(u) \rangle| \, dt.$$

Since $\|\alpha_j^{(i)}(u)\| \leq 1$ and since $\alpha_j^{(i)}$ has compact support, it follows from the definition of the norm of a Radon measure that the integral on the right in the last inequality is less than

$$\sum_{j=1}^{p_i} \int_R \|\mu_t(\sigma) - \mu_t(\hat{\sigma})\| \, dt = p_i \int_R \|\mu_t(\sigma) - \mu_t(\hat{\sigma})\| \, dt.$$

The Approximation Lemma

Since by assumption the family $\mu_t(\sigma)$ depends strongly continuously on $\sigma \in \Sigma$ we have that

$$p_i \int_R \|\mu_t(\sigma) - \mu_t(\hat{\sigma})\| \, dt \to 0 \qquad (\sigma \to \hat{\sigma}).$$

Thus (3.2) is established.

For every $i = 1, 2, \ldots$, we choose an arbitrary point $u_j^{(i)}$ in the intersection $O_j^{(i)} \cap N$, $j = 1, \ldots, p_i$. If the intersection is empty, then we set $u_j^{(i)} = \hat{u}$, where $\hat{u}$ is a fixed point in N. We obtain the sets of points

$$\{u_1^{(i)}, \ldots, u_{p_i}^{(i)}\} \subset N, \qquad i = 1, 2, \ldots.$$

If we assume that the functions $\alpha_j^{(i)}(u)$ in the formula (3.1) are sufficiently "close" to the characteristic functions of the sets $O_j^{(i)}$, then the corresponding functions $\lambda_j^{(i)}(t; \sigma)$ express approximately the parts of the unit measure $\mu_t(\sigma)$ which are concentrated on the sets $O_j^{(i)} \cap N$. For this reason, we can assume for large i, when the diameters of all the sets $O_j^{(i)}$ are small, that the entire measure $\mu_t(\sigma)$ is concentrated at the points $u_1^{(i)}, \ldots, u_{p_i}^{(i)}$ in the quantities $\lambda_1^{(i)}(t, \sigma), \ldots, \lambda_{p_i}^{(i)}(t, \sigma)$. In so doing, we make an error (in the sense of weak convergence) which becomes smaller as i becomes larger.

We can express this intuitively obvious fact in a precise way, saying that the sequence of generalized controls (which are chattering controls)

$$\Delta_t^{(i)}(\sigma) = \sum_{j=1}^{p_i} \lambda_j^{(i)}(t; \sigma) \delta u_i^{(j)} \tag{3.3}$$

converges weakly to the generalized control $\mu_t(\sigma)$ as $i \to \infty$, uniformly with respect to $\sigma \in \Sigma$.

In order to prove this, we take a continuous function $g(t, u)$ with a compact support, and let I_g be the projection of the support onto the t-axis. Furthermore, we use the notation

$$\eta^{(i)} = \max_j \sup_{(t,u)} \{|g(t, u) - g(t, u_j^{(i)})| : (t, u) \in I_g \times O_j^{(i)}, j = 1, \ldots, p_i\}.$$

Obviously,

$$\eta^{(i)} \to 0 \qquad (i \to \infty).$$

Therefore,

$$\left| \int_R \langle \mu_t(\sigma) - \Delta_t^{(i)}(\sigma), g(t, u) \rangle \, dt \right|$$

$$= \left| \int_R \langle \mu_t(\sigma), \sum_{j=1}^{p_i} \alpha_j^{(i)}(u) g(t, u) \rangle \, dt - \sum_{j=1}^{p_i} \int_R \lambda_j^{(i)}(t; \sigma) g(t, u_j^{(i)}) \, dt \right|$$

$$= \left| \int_R \left\langle \mu_t(\sigma), \sum_{j=1}^{p_i} \alpha_j^{(i)}(u)[g(t, u) - g(t, u_j^{(i)})] \right\rangle dt \right|$$

$$\leq \sum_{j=1}^{p_i} \int_{I_g} \left\langle \mu_t(\sigma), \alpha_j^{(i)}(u) |g(t, u) - g(t, u_j^{(i)})| \right\rangle dt$$

$$\leq \eta^{(i)} \sum_{j=1}^{p_i} \int_{I_g} \left\langle \mu_t(\sigma), \alpha_j^{(i)}(u) \right\rangle dt = \eta^{(i)} \int_{I_g} dt \int_N \sum_{j=1}^{p_i} \alpha_j^{(i)}(u) \, d\mu_t(\sigma)$$

$$= \eta^{(i)} \int_{I_g} dt \int_N d\mu_t(\sigma) = \eta^{(i)} \int_{I_g} dt = \eta^{(i)} |I_g| \to 0 \quad (i \to \infty),$$

where, as is seen from the estimate, the convergence is uniform with respect to $\sigma \in \Sigma$.

We now construct a sequence of piecewise-constant controls

$$\delta_{u^{(i)}(t;\sigma)} \in \Omega_U, \quad i = 1, 2, \ldots,$$

which depends on σ strongly continuously for every $i = 1, 2, \ldots$ and is such that we have

$$\int_R \left\langle \Delta_t^{(i)}(\sigma) - \delta_{u^{(i)}(t;\sigma)}, g(t, u) \right\rangle dt \to 0 \quad (i \to \infty), \tag{3.4}$$

uniformly with respect to $\sigma \in \Sigma$, for any continuous function $g(t, u)$ with compact support.

For every $i = 1, 2, \ldots$, we partition the interval $[-i, i]$ into $2i^2$ intervals $I_k^{(i)}$ of the same length $1/i$, $k = 1, 2, \ldots, 2i^2$. Further, we partition every interval $I_k^{(i)}$ into p_i subintervals $I_{kj}^{(i)}(\sigma)$, $j = 1, \ldots, p_i$ (which depend on $\sigma \in \Sigma$) of length

$$|I_{kj}^{(i)}(\sigma)| = \int_{I_k^{(i)}} \lambda_j^{(i)}(t; \sigma) \, dt. \tag{3.5}$$

This can be done, since $\sum_{j=1}^{p_i} \lambda_j^{(i)}(t; \sigma) \equiv 1$ for all $\sigma \in \Sigma$, $i = 1, 2, \ldots$.

We assume that the intervals $I_{kj}^{(i)}(\sigma)$ are enumerated with index j so that the interval $I_{kj'}^{(i)}(\sigma)$ lies to the left of the interval $I_{kj''}^{(i)}(\sigma)$ if $j' \leq j''$. If $\lambda_j^{(i)}(t; \sigma) = 0$ for all $t \in I_k^{(i)}$, then the corresponding interval $I_{kj}^{(i)}$ degenerates into a point.

We define the function $u^{(i)}(t, \sigma)$ (to within its values on a finite number of

The Approximation Lemma

points), setting it equal to $\hat{u}$ outside of the interval $[-i, i]$, and equal to $u_j^{(i)}$ on each of the intervals $I_{kj}^{(i)}(\sigma)$:

$$u^{(i)}(t; \sigma) = \hat{u} \quad \forall\, t \notin [-i, i],$$
$$u^{(i)}(t; \sigma) = u_j^{(i)} \quad \forall\, t \in I_{kj}^{(i)} \quad \text{and} \quad \forall\, k = 1, 2, \ldots, 2i^2. \quad (3.6)$$

In order to prove statement (3.4), we assume that i is so large that the inclusion $I_g \subset [-i, i]$ holds (as before, I_g is the projection of the support of $g(t, u)$ onto the t-axis), and we choose an arbitrary point $t_k^{(i)}$ in every interval $I_k^{(i)}$. Then

$$\int_R \langle \Delta_t^{(i)}(\sigma) - \delta_{u^{(i)}(t;\sigma)},\, g(t, u) \rangle\, dt = \sum_{k=1}^{2i^2} \int_{I_k^{(i)}} \langle \Delta_t^{(i)}(\sigma) - \delta_{u^{(i)}(t;\sigma)},\, g(t_k^{(i)}, u) \rangle\, dt$$

$$+ \sum_{k=1}^{2i^2} \int_{I_k^{(i)}} \langle \Delta_t^{(i)}(\sigma) - \delta_{u^{(i)}(t;\sigma)},\, g(t, u) - g(t_k^{(i)}, u) \rangle\, dt$$

$$= \sum_{k=1}^{2i^2} \int_{I_k^{(i)}} \langle \Delta_t^{(i)}(\sigma) - \delta_{u^{(i)}(t;\sigma)},\, g(t, u) - g(t_k^{(i)}, u) \rangle\, dt,$$

because by virtue of (3.3), (3.6), and (3.5), we have

$$\sum_{k=1}^{2i^2} \int_{I_k^{(i)}} \langle \Delta_t^{(i)}(\sigma) - \delta_{u^{(i)}(t,\sigma)},\, g(t_k^{(i)}, u) \rangle\, dt$$

$$= \sum_{k=1}^{2i^2} \left\{ \int_{I_k^{(i)}} \sum_{j=1}^{pi} \lambda_j^{(i)}(t; \sigma) g(t_k^{(i)}, u_j^{(i)})\, dt - \sum_{j=1}^{p} \int_{I_{kj}^{(i)}} g(t_k^{(i)}, u_j^{(i)})\, dt \right\}$$

$$= \sum_{k=1}^{2i^2} \left\{ \sum_{j=1}^{pi} g(t_k^{(i)}, u_j^{(i)}) \left(\int_{I_k^{(i)}} \lambda_j^{(i)}(t; \sigma)\, dt - \int_{I_{kj}^{(i)}} dt \right) \right\} = 0.$$

Therefore,

$$\left| \int_R \langle \Delta_t^{(i)}(\sigma) - \delta_{u^{(i)}(t;\sigma)},\, g(t, u) \rangle\, dt \right| \leq \sum_{k=1}^{2i^2} \int_{I_k^{(i)}} |\langle \Delta_t^{(i)}(\sigma) - \delta_{u^{(i)}(t;\sigma)},\, g(t, u) - g(t_k^{(i)}, u) \rangle|\, dt.$$

We shall estimate the expression on the right-hand side.

Let I be an arbitrary interval of the t-axis whose interior contains the compact set I_g. Then, for i sufficiently large,

$$\sum_{k=1}^{2i^2} \int_{I_k^{(i)}} |\langle \Delta_t^{(i)}(\sigma) - \delta_{u^{(i)}(t;\sigma)}, g(t, u) - g(t_k^{(i)}, u) \rangle| \, dt$$

$$\leq \sum_{I_k^{(i)} \subset I} \int_{I_k^{(i)}} |\langle \Delta_t^{(i)}(\sigma) - \delta_{u^{(i)}(t;\sigma)}, g(t, u) - g(t_k^{(i)}, u) \rangle| \, dt,$$

where the summation on the right-hand side is carried, as indicated, only over the intervals $I_k^{(i)}$ which are contained in I. The length of $I_k^{(i)}$ is $1/i$. The measure $\Delta_t^{(i)}(\sigma) - \delta_{u^{(i)}(t;\sigma)}$ is concentrated on the bounded set N, and its norm is no greater than 2:

$$\|\Delta_t^{(i)}(\sigma) - \delta_{u^{(i)}(t;\sigma)}\| \leq \left\| \sum_{j=1}^{p_i} \lambda_j^{(i)}(t;\sigma) \delta_{u_j^{(i)}} \right\| + \|\delta_{u^{(i)}(t;\sigma)}\| \leq \sum_{j=1}^{p_i} \lambda_j^{(i)}(t;\sigma) + 1 = 2.$$

Therefore, using the notation

$$\zeta^{(i)} = \sup \{|g(t', u) - g(t'', u)|: u \in N, t', t'' \in I, |t' - t''| \leq 1/i\},$$

we have the relation

$$\sum_{I_k^{(i)} \subset I} \int_{I_k^{(i)}} |\langle \Delta_t^{(i)}(\sigma) - \delta_{u^{(i)}(t;\sigma)}, g(t, u) - g(t_k^{(i)}, u) \rangle| \, dt$$

$$\leq \int_I 2\zeta^{(i)} \, dt = 2|I|\zeta^{(i)} \to 0 \quad (i \to \infty),$$

which proves formula (3.4).

In order to finish the proof of the theorem, it remains to prove the strong continuity of the family of controls $\delta_{u^{(i)}(t;\sigma)}$. To this end, we note that, if $\hat{t}$ is an arbitrary interior point of the interval $I_{kj}^{(i)}(\hat{\sigma})$, and if the point σ is sufficiently close to $\hat{\sigma}$, then, by (3.5) and (3.2), $\hat{t}$ is also an interior point of $I_{kj}^{(i)}(\sigma)$. Also, by the definition of functions $u^{(i)}(t;\sigma)$, we obtain

$$u^{(i)}(\hat{t}; \sigma) = u^{(i)}(\hat{t}; \hat{\sigma}) = u_j^{(i)}.$$

Therefore, the norm of the difference $\|\delta_{u^{(i)}(t;\sigma)} - \delta_{u^{(i)}(t;\hat{\sigma})}\|$ tends to zero as $\sigma \to \hat{\sigma}$ at all points $t \in R$ which are distinct from the end points of the intervals $I_{kj}^{(i)}(\hat{\sigma})$. Thus, applying the Lebesgue theorem on passing to the limit under the integral, we obtain the desired strong continuity:

$$\int_R \|\delta_{u^{(i)}(t;\sigma)} - \delta_{u^{(i)}(t;\hat{\sigma})}\| \, dt \to 0 \quad (\sigma \to \hat{\sigma}) \quad \forall \ i = 1, 2, \ldots.$$

Remark on the Terminology. It can be seen from the proof we presented that, if a generalized control

$$\Delta_t = \sum_{j=1}^{p} \lambda_j(t)\delta_{u_j}, \quad \sum_{j=1}^{p} \lambda_j(t) \equiv 1, \quad u_j \in U, j=1,\ldots,p, \quad (3.7)$$

is given, then our process of constructing piecewise-continuous controls (3.6) yields a sequence of functions $u^{(i)}(t)$ whose values rapidly (with frequency i) oscillate on the interval $-i \leq t \leq i$ among the values $u_1, \ldots, u_p$. That is, the function $u^{(i)}(t)$ remains on each of the intervals

$$\frac{k}{i} \leq t \leq \frac{k+1}{i}, \quad k = -i^2, \ldots, i^2,$$

at the point u_j for the time

$$\int_{[k/i,(k+1)/i]} \lambda_j(t)\, dt, \quad j=1,\ldots,p.$$

This sequence of functions which oscillate faster and faster converges weakly to the generalized control Δ_t. This is why we said in Chapter 2 that the controls Δ_t "chatter."

Actually, we used the term *chattering controls* for more general controls where the points u_j are replaced by arbitrary bounded measurable functions of t which take on values in U. Also in this case, an insignificant change in the process allows us to construct a sequence of piecewise-constant controls that converges weakly to Δ_t. In a sense, these controls can also be assumed to oscillate rapidly among—now variable—points $u_1(t), \ldots, u_p(t)$ with frequency i for the ith function of the sequence.

Assume that piecewise-constant controls approximate an arbitrary generalized control μ_t, and not a chattering control (3.7). Then, as can be seen from the proof, the approximation becomes better as the number of points among which the approximating function oscillates becomes larger. "In the limit," the function is "distributed" over the entire set U with the required density in the form of the measure μ_t. If μ_t has the form (3.7), then this limit "distribution" is accomplished among p points $u_1, \ldots, u_p$ with the densities which coincide with the functions $\lambda_1(t), \ldots, \lambda_p(t)$ for almost all t.

4

The Existence and Continuous Dependence Theorem for Solutions of Differential Equations

The basic result of this chapter is formulated in Theorems 4.3 and 4.4 on the existence and continuous dependence of solutions of differential equations on the initial data and right-hand sides. These theorems are formulated and proved in Sections 4.4 and 4.5, and they appear as comparatively easy corollaries of Theorem 4.2 on the existence and continuous dependence of solutions of differential equations with fixed right-hand sides, which is proved in Section 4.3. The technique used for deriving Theorems 4.3 and 4.4 from Theorem 4.2 is standard, and it often turns out to be very useful.

In Sections 4.1 and 4.2, we gathered all auxiliary material and presented a fixed point theorem for contraction mappings in a convenient form for the proof of Theorem 4.2.

4.1. Preparatory Material

We denote by E_{Lip} the linear space of n-dimensional functions $F(t, x)$ which are defined on $R \times R^n$ and satisfy the following conditions: Every function $F(t, x) \in E_{\text{Lip}}$ has a compact support, is (Lebesgue) measurable in $t \in R$ for fixed $x \in R^n$, and is majorized by a function $m_F(t)$ summable on R:

$$|F(t, x)| \leqslant m_F(t) \quad \forall (t, x) \in R \times R^n, \qquad \int_R m_F(t)\, dt < \infty. \qquad (4.1)$$

Moreover, we assume that the functions $F(t, x)$ are *Lipschitzian* in x: There exists a function $L_F(t)$ which is finite for all t in R, which is summable on R, and is such that

$$|F(t, x') - F(t, x'')| \leq L_F(t)|x' - x''| \quad \forall\, (t, x'), (t, x'') \in R \times R^n. \quad (4.2)$$

It follows directly from the fact that $F(t, x)$ is Lipschitzian that it is also continuous in x for any fixed t. We shall show that the continuity in x implies, in turn, the (Lebesgue) measurability of the function $F(t, x(t))$, $t \in R$, for any n-dimensional (Lebesgue) measurable function $x(t)$, $t \in R$.

The assertion is obvious if $x(t)$ takes on a finite number of values $x_1, \ldots, x_p$ which can be assumed to be distinct. In this case,

$$x(t) = \sum_{i=1}^{p} \chi_i(t) x_i,$$

where $\chi_i(t)$ are the characteristic functions of the subsets of R (Lebesgue measurable and mutually disjoint) on which $x(t)$ takes on the values x_i, respectively. Thus, the measurability of the function

$$F(t, x(t)) = F\left(t, \sum_{i=1}^{p} \chi_i(t) x_i\right) = \sum_{i=1}^{p} \chi_i(t) F(t, x_i), \quad t \in R,$$

follows from the measurability of the functions $F(t, x_i)$, $i = 1, \ldots, p$.

An arbitrary measurable function $x(t)$ with $t \in R$ can be represented as a pointwise limit of the sequence of measurable functions

$$x_1(t), x_2(t), \ldots, \quad t \in R,$$

which take on a finite number of values. Since $F(t, x)$ is continuous in x, we have

$$F(t, x(t)) = \lim_{j \to \infty} F(t, x_j(t)) \quad \forall\, t \in R.$$

Therefore, the function $F(t, x(t))$ is measurable, as it is a pointwise limit of measurable functions.

Moreover, from the estimate (4.1) follows the summability of the function $F(t, x(t))$ on R. Therefore, the indefinite integral

$$\int_{t_1}^{t} F(\theta, x(\theta))\, d\theta, \quad t \in R,$$

has sense for an arbitrary measurable function $x(t)$, $t \in R$.

Solutions of Differential Equations: Existence

We consider the differential equation

$$\dot{x} = F(t, x), \qquad F(t, x) \in E_{\text{Lip}}. \tag{4.3}$$

Any absolutely continuous function $x(t)$ which is defined on the entire t-axis and satisfies the equality

$$\dot{x}(t) = F(t, x(t))$$

for almost all $t \in R$ is said to be a *solution* of this equation. Integrating both parts of the last equality from τ up to t, and taking into account the absolute continuity of the function $x(t)$, we obtain the following integral equation in the unknown function $x(t)$, $t \in R$,

$$x(t) = x_\tau + \int_\tau^t F(\theta, x(\theta)) \, d\theta, \qquad x_\tau = x(\tau),$$

which is equivalent to the differential equation (4.3) together with the *initial condition* for the solution $x(t)$

$$x(\tau) = x_\tau.$$

Without the condition that the required function $x(t)$ be absolutely continuous we could not have accomplished the transfer from the differential equation to the integral equation, since the integral of the derivative $\dot{x}(t)$ from τ to t may not be equal to the difference $x(t) - x(\tau)$. This is well known from the theory of functions (let us recall the singular Cantor function).

Let us note that any solution of equation (4.3) takes on constant values outside of some interval $[t_1, t_2]$; it is sufficient that the interval $[t_1, t_2]$ contain the projection of the support of $F(t, x)$ onto the t-axis.

In order to formulate Theorem 4.2 on the existence and continuous dependence of solutions of the differential equation (4.3) on the initial data and right-hand side, we must introduce a seminorm $\|\cdot\|_w$ on E_{Lip}. With the aid of this seminorm, we shall estimate the distance between any two functions in E_{Lip}. We must also define the notion of a uniformly Lipschitzian subset of E_{Lip}.

We begin with the definition of the seminorm $\|\cdot\|_w$. First, we shall prove that, for any function $F(t, x) \in E_{\text{Lip}}$, the function

$$g(t', t'', x) = \int_{t'}^{t''} F(t, x) \, dt, \qquad (t', t'', x) \in R \times R \times R^n,$$

is jointly continuous in all of its arguments. Indeed,

$$|g(t'+\delta t', t''+\delta t'', x+\delta x)-g(t', t'', x)|$$

$$\leq \int_{t'-|\delta t'|}^{t'+|\delta t'|} |F(t, x+\delta x)|\, dt + \int_{t''-|\delta t''|}^{t''+|\delta t''|} |F(t, x+\delta x)|\, dt + \left|\int_{t'}^{t''} |F(t, x+\delta x) - F(t, x)|\, dt\right|$$

$$\leq \int_{t'-|\delta t'|}^{t'+|\delta t'|} m_F(t)\, dt + \int_{t''-|\delta t''|}^{t''+|\delta t''|} m_F(t)\, dt + |\delta x| \int_R L_F(t)\, dt \to 0$$

as $\delta t', \delta t'', \delta x \to 0$. We now define the seminorm $\|\cdot\|_w$ by the formula

$$\|F(t, x)\|_w = \max_{t', t'', x} \left|\int_{t'}^{t''} F(t, x)\, dt\right|,$$

where the integration takes place for a fixed x, and the maximum is taken over all $t', t'' \in R$, $x \in R^n$.*

A sequence of functions $F_j(t, x), j = 1, 2, \ldots$ converges to $F(t, x)$ in the topology determined by the seminorm $\|\cdot\|_w$ if and only if

$$\|F_j(t, x) - F(t, x)\|_w \to 0 \qquad (j \to \infty).$$

The topology defined by this seminorm is very "coarse": A function $F(t, x)$ can have an arbitrarily small seminorm $\|F(t, x)\|_w$, i.e., it can belong to an arbitrarily small neighborhood of zero in the topology under consideration and, nevertheless, this function can have large absolute values on an arbitrary given cube in the space $R \times R^n$. A typical example of a function that is small in the sense of this seminorm is a function that rapidly oscillates in t, e.g., the function

$$F(t, x) = g(t, x) \cos pt$$

for large p. Here, $g(t, x)$ is a function which has a compact support and is continuously differentiable with respect to all its arguments. In order to

---

*The number $\|\cdot\|_w$ is, obviously, a seminorm:

$$\|F_w\| \geq 0, \qquad \|\lambda F\|_w = |\lambda|\, \|F\|_w, \qquad \|F_1 + F_2\|_w \leq \|F_1\|_w + \|F_2\|_w.$$

It becomes a norm if any two functions whose difference is zero for every fixed x and for almost all $t \in R$ are identified in E_{Lip}. The function $\rho_w(F_1, F_2) = \|F_1 - F_2\|_w$ is a pseudometric in E_{Lip}, and it becomes a metric after this identification.

Solutions of Differential Equations: Existence

estimate $\|g(t, x) \cos pt\|_w$, we carry out the following integration by parts:

$$\int_{t'}^{t''} g(t, x) \cos pt \, dt = g(t'', x) \int_{t'}^{t''} \cos pt \, dt - \int_{t'}^{t''} \left(\int_{t'}^{t} \cos p\theta \, d\theta \right) \left(\frac{\partial}{\partial t} g(t, x) \right) dt$$

$$= \frac{1}{p} \left\{ (\sin pt'' - \sin pt') g(t'', x) - \int_{t'}^{t''} (\sin pt - \sin pt') \frac{\partial}{\partial t} g(t, x) \, dt \right\}.$$

Hence,

$$\left| \int_{t'}^{t''} g(t, x) \cos pt \, dt \right| \leq \frac{1}{p} 2 \left\{ |g(t'', x)| + \left| \int_{t'}^{t''} \left| \frac{\partial}{\partial t} g(t, x) \right| dt \right| \right\} \leq \frac{\text{const}}{p} \to 0 \quad (p \to \infty).$$

We shall consider the linear space

$$E_\sigma = R \times R^n \times E_{\text{Lip}} = \{\sigma = (\tau, x_\tau, F(t, x)) : (\tau, x_\tau) \in R \times R^n, F(t, x) \in E_{\text{Lip}}\}$$

with the seminorm

$$\|\sigma\|_w = \|(\tau, x_\tau, F(t, x))\|_w = |\tau| + |x_\tau| + \|F(t, x)\|_w$$

as the product of the space of *initial data* $(\tau, x_\tau) \in R \times R^n$ for the solution of equation (4.3) with the space of the right-hand side $F(t, x) \in E_{\text{Lip}}$.

The solution $x(t)$, $t \in R$, of the differential equation (4.3) that satisfies the initial condition $x(\tau) = x_\tau$ will be written in the form

$$x(t) = x(t; \tau, x_\tau, F) = x(t; \sigma), \quad t \in R.$$

We shall say that this solution *depends continuously on the initial data and right-hand side when the parameter* $\sigma = (\tau, x_\tau, F(t, x))$ ranges over a subset $M_\sigma \subset E_\sigma$ if, for every $\hat{\sigma} = (\hat{\tau}, \hat{x}_{\hat{\tau}}, \hat{F}) \in M_\sigma$ and every $\varepsilon > 0$, there exists a $\delta > 0$ such that the relation

$$\|\sigma - \hat{\sigma}\|_w = |\tau - \hat{\tau}| + |x_\tau - \hat{x}_{\hat{\tau}}| + \|F(t, x) - \hat{F}(t, x)\|_w \leq \delta,$$

$$\sigma = (\tau, x_\tau, F(t, x)) \in M_\sigma$$

implies

$$\max_{t \in R} |x(t; \tau, x_\tau, F) - x(t; \hat{\tau}, \hat{x}_{\hat{\tau}}, \hat{F})| = \|x(t; \sigma) - x(t; \hat{\sigma})\| \leq \varepsilon.$$

A set $M \subset E_{\text{Lip}}$ will be called *uniformly Lipschitzian* if, for any $F(t, x) \in M$, there exists a function $L_F(t)$ which is summable on R and such that

$$|F(t, x') - F(t, x'')| \leq L_F(t)|x' - x''|, \quad \int_R L_F(t)\, dt \leq C, \qquad (4.4)$$

where the constant C does not depend on the choice of $F(t, x) \in M$.

The following assertion expresses a property of uniformly Lipschitzian sets that is of importance to us:

Assertion 4.1. Let Φ be an equicontinuous family of n-dimensional functions $x(t)$, $t \in R$, which take on constant values outside of a fixed interval $[T_1, T_2]$ not dependent on $x(t) \in \Phi$, and let M be an arbitrary uniformly Lipschitzian subset of E_{Lip}. Then, for every $\varepsilon > 0$, there exists a positive number $\delta > 0$ such that the conditions

$$\|F(t, x)\|_w \leq \delta \quad \text{and} \quad F(t, x) \in M$$

imply

$$\max_{t', t'' \in R} \left| \int_{t'}^{t''} F(t, x(t))\, dt \right| \leq \varepsilon \quad \forall\, x(t) \in \Phi.$$

Proof. It follows from the equicontinuity of the family Φ that there exists a function $\omega(t)$, $t \geq 0$, and a positive number $\eta_0 > 0$ such that

(i) $\omega(|t|) \to 0$ $(t \to 0)$ and $\omega(0) = 0$;

(ii) $|x(t') - x(t'')| \leq \omega(|t' - t''|) \quad \forall x(t) \in \Phi,$ for $|t' - t''| \leq \eta_0$.

For any interval $[t_1, t_2]$ and a number $\eta > 0$, the intersection $[t_1, t_2] \cap [T_1, T_2]$ can be partitioned into

$$p = \frac{T_2 - T_1}{\eta} + 1$$

subintervals of length no greater than η. Therefore, if $\eta \leq \eta_0$, then the oscillation of any function $x(t) \in \Phi$ does not exceed $\omega(\eta)$ on each of these intervals. By assumption, the oscillation of $x(t)$ is zero on the parts of the interval $[t_1, t_2]$ that lie to the right or left of the intersection (if there are any such parts). It follows that, for any $\eta \in [0, \eta_0]$, an arbitrary interval $[t_1, t_2]$ of the time axis can be partitioned into

$$p \leq \frac{T_2 - T_1}{\eta} + 3$$

Solutions of Differential Equations: Existence

subintervals on each of which the oscillation of any function $x(t) \in \Phi$ does not exceed $\omega(\eta)$.

Assume that, for all $F(t, x) \in M$ and all $x(t) \in \Phi$, we have

$$\max_{t', t'' \in R} \left| \int_{t'}^{t''} F(t, x(t)) \, dt \right| = \left| \int_{t_1}^{t_2} F(t, x(t)) \, dt \right|,$$

where t_1 and t_2 are the points at which the maximum on the left-hand side is attained. We partition the interval $[t_1, t_2]$ in the way indicated above into

$$p \leq \frac{T_2 - T_1}{\eta} + 3, \qquad 0 < \eta < \eta_0,$$

subintervals by the partition points

$$t_1 = \theta_0 \leq \theta_1 \leq \cdots \leq \theta_{p-1} \leq \theta_p = t_2.$$

Then

$$\left| \int_{t_1}^{t_2} F(t, x(t)) \, dt \right| \leq \sum_{i=0}^{p-1} \int_{\theta_i}^{\theta_{i+1}} |F(t, x(t)) - F(t, x(\theta_i))| \, dt$$

$$+ \sum_{i=0}^{p-1} \left| \int_{\theta_i}^{\theta_{i+1}} F(t, x(\theta_i)) \, dt \right|$$

$$\leq \sum_{i=0}^{p-1} \int_{\theta_i}^{\theta_{i+1}} L_F(t) |x(t) - x(\theta_i)| \, dt + p \|F(t, x)\|_w$$

$$\leq C\omega(\eta) + \left(\frac{T_2 - T_1}{\eta} + 3 \right) \|F(t, x)\|_w,$$

where C is the constant of (4.4) for the set M under consideration.

If $\|F(t, x)\|_w = 0$, then

$$\left| \int_{t_1}^{t_2} F(t, x(t)) \, dt \right| \leq C\omega(\eta) \qquad \forall \eta \in [0, \eta_0].$$

Thus, the left-hand side is zero, since $\omega(\eta) \to 0$ $(\eta \to 0)$. On the other hand, if $0 < \|F(t, x)\|_w \leq \eta_0^2$, then, taking

$$\eta = (\|F(t, x)\|_w)^{1/2} \leq \eta_0,$$

we obtain

$$\left|\int_{t_1}^{t_2} F(t, x(t))\, dt\right| \leq C\omega(\|F(t, x)\|_w)^{1/2} + (T_2 - T_1)(\|F(t, x)\|_w)^{1/2} + 3\|F(t, x)\|_w.$$

Since the right-hand side of the inequality thus obtained tends to zero together with $\|F(t, x)\|_w$, we have that, for every $\varepsilon > 0$, there exists an $\eta_1 > 0$ such that the relation $\|F(t, x)\|_w \leq \min(\eta_0^2, \eta_1)$ implies

$$\max \left|\int_{t'}^{t''} F(t, x(t))\, dt\right| \leq \varepsilon \qquad \forall x(t) \in \Phi.$$

Therefore, we can set $\delta = \min(\eta_0^2, \eta_1)$.

4.2. A Fixed-Point Theorem for Contraction Mappings

Let X be a metric space, let ρ be a distance function on X, and let φ be an arbitrary (not necessarily continuous) mapping of the space X into itself.

We define the *iterations* of the mapping φ by induction:

$$\varphi^{(1)} = \varphi, \qquad \varphi^{(p)} = \varphi \circ \varphi^{(p-1)} = \varphi^{(p-1)} \circ \varphi.$$

Therefore,

$$\varphi^{(1)}(x) = \varphi(x), \qquad \varphi^{(p)}(x) = \varphi(\varphi^{(p-1)}(x)) = \varphi^{(p-1)}(\varphi(x)) \qquad \forall x \in X.$$

The mapping φ is said to be a *contraction mapping (contraction)* if there exists a positive number $k < 1$, called the *contraction coefficient*, such that

$$\rho(\varphi(x), \varphi(y)) \leq k\rho(x, y) \qquad \forall x, y \in X.$$

Let Σ be an arbitrary topological space of points σ. The family of mappings $\varphi(x, \sigma)$ of the space X into itself which depends on the parameter $\sigma \in \Sigma$,

$$\varphi(\cdot, \sigma) \colon X \to X, \qquad \sigma \in \Sigma,$$

will be said to *depend continuously on* $\sigma \in \Sigma$ if, for every *fixed* $x \in X$, the mapping

$$\varphi(x, \cdot) \colon \Sigma \to X$$

is continuous. This family will be called a *uniform contraction* family of

Solutions of Differential Equations: Existence 61

mappings of X into itself if there exists a positive number $k<1$ which *does not depend* on σ and is such that, for every fixed $\sigma \in \Sigma$, the mapping

$$\varphi(\cdot, \sigma): X \to X$$

is a contraction with coefficient k:

$$\rho(\varphi(x, \sigma), \varphi(y, \sigma)) \leqslant k\rho(x, y) \qquad \forall x, y \in X \text{ and } \forall \sigma \in \Sigma.$$

Theorem 4.1. Let X be a complete metric space, let Σ be a topological space, and let $\varphi(x, \sigma)$ with $\sigma \in \Sigma$ be an arbitrary family of mappings of X into itself. We assume that some iteration of the family

$$\varphi^{(p)}(\cdot, \sigma): X \to X, \qquad \sigma \in \Sigma,$$

depends continuously on $\sigma \in \Sigma$ and is a uniform contraction family of mappings of X into itself with a contraction coefficient $k<1$. Then every mapping from the initial family

$$\varphi(\cdot, \sigma): X \to X, \qquad \sigma \in \Sigma,$$

has a unique fixed point x_σ,

$$\varphi(x_\sigma, \sigma) = x_\sigma,$$

which depends continuously on the parameter σ, i.e., the mapping $\sigma \to x_\sigma$ of the topological space Σ into X is continuous.

Proof. The theorem on the existence of a fixed point of a contraction mapping in its standard form guarantees the existence of a unique fixed point x_σ of the iteration $\varphi^{(p)}$ for every $\sigma \in \Sigma$:

$$\varphi^{(p)}(x_\sigma, \sigma) = x_\sigma.$$

Applying the mapping $\varphi(\cdot, \sigma)$ to both parts of this equality, we obtain the equality

$$\varphi(\varphi^{(p)}(x_\sigma, \sigma), \sigma) = \varphi^{(p)}(\varphi(x_\sigma, \sigma), \sigma) = \varphi(x_\sigma, \sigma).$$

which shows that, together with x_σ, the point $\varphi(x_\sigma, \sigma)$ is also a fixed point of the mapping $\varphi^{(p)}(\cdot, \sigma)$. Therefore, we obtain from the uniqueness of the fixed point that

$$\varphi(x_\sigma, \sigma) = x_\sigma,$$

i.e., x_σ is a fixed point of the mapping $\varphi(\cdot, \sigma)$. The mapping $\varphi(\cdot, \sigma)$ has no other fixed points, since every fixed point of any mapping is also a fixed point of any iteration of this mapping, and the iteration $\varphi^{(p)}$ has a unique fixed point for every $\sigma \in \Sigma$.

In order to prove the continuous dependence of x_σ on σ, we write the following obvious inequalities:

$$\rho(x_\sigma, x_{\hat\sigma}) = \rho(\varphi^{(p)}(x_\sigma, \sigma), \varphi^{(p)}(x_{\hat\sigma}, \hat\sigma))$$
$$\leq \rho(\varphi^{(p)}(x_\sigma, \sigma), \varphi^{(p)}(x_{\hat\sigma}, \sigma)) + \rho(\varphi^{(p)}(x_{\hat\sigma}, \sigma), \varphi^{(p)}(x_{\hat\sigma}, \hat\sigma))$$
$$\leq k\rho(x_\sigma, x_{\hat\sigma}) + \rho(\varphi^{(p)}(x_{\hat\sigma}, \sigma), \varphi^{(p)}(x_{\hat\sigma}, \hat\sigma)).$$

Hence,

$$\rho(x_\sigma, x_{\hat\sigma}) \leq \frac{1}{1-k} \rho(\varphi^{(p)}(x_{\hat\sigma}, \sigma), \varphi^{(p)}(x_{\hat\sigma}, \hat\sigma)).$$

It follows from the continuous dependence of $\varphi^{(p)}$ on the parameter that, for every $\varepsilon > 0$, there exists a neighborhood $V_{\hat\sigma}$ of a point $\hat\sigma$ such that for $\sigma \in V_{\hat\sigma}$ we have

$$\rho(\varphi^{(p)}(x_{\hat\sigma}, \sigma), \varphi^{(p)}(x_{\hat\sigma}, \hat\sigma)) \leq \varepsilon(1-k).$$

Therefore, we obtain the inequality

$$\rho(x_\sigma, x_{\hat\sigma}) \leq \varepsilon \quad \forall \, \sigma \in V_{\hat\sigma},$$

which proves the continuity of x_σ at an arbitrary point $\hat\sigma \in \Sigma$.

In conclusion, we note that any fixed point x_σ can be obtained by the successive approximation method: If, beginning with an arbitrary point $x_0 \in X$, we construct the sequence

$$x_0, \quad x_1 = \varphi(x_0, \sigma), \quad x_2 = \varphi(x_1, \sigma) = \varphi^{(2)}(x_0, \sigma), \ldots, x_{i+1}$$
$$= \varphi(x_i, \sigma) = \varphi^{(i+1)}(x_0, \sigma), \ldots,$$

then

$$x_i = \varphi^{(i)}(x_0, \sigma) \to x_\sigma \quad (i \to \infty).$$

In this connection, it is of importance to keep in mind that a sequence of approximations $x_0, x_1, \ldots$ is constructed with the aid of the initial (possibly not contraction) mapping $\varphi(\cdot, \sigma)$, and not with the aid of one of its iterations $\varphi^{(p)}(\cdot, \sigma)$. In order to prove this assertion, we note that each of the p sequences

$$x_0, \quad x_p = \varphi^{(p)}(x_0, \sigma), \ldots, \quad x_{ip} = \varphi^{(p)}(x_{(i-1)p}, \sigma),$$
$$x_1, \quad x_{p+1} = \varphi^{(p)}(x_1, \sigma), \ldots, \quad x_{ip+1} = \varphi^{(p)}(x_{(i-1)p+1}, \sigma),$$
$$x_{p-1}, \quad x_{2p-1} = \varphi^{(p)}(x_{p-1}, \sigma), \ldots, \quad x_{ip+p-1} = \varphi^{(p)}(x_{(i-1)p+p-1}, \sigma), \ldots$$

Solutions of Differential Equations: Existence

converges, according to the standard fixed-point theorem, to the same limit x_σ. Hence $x_i \to x_\sigma$ $(i \to \infty)$.

4.3. The Existence and Continuous Dependence Theorem for Solutions of Equation (4.3)

Theorem 4.2. Any equation

$$\dot{x} = F(t, x), \qquad F(t, x) \in E_{\text{Lip}},$$

has a unique solution

$$x(t) = x(t; \tau, x_\tau, F), \qquad t \in R,$$

that satisfies an arbitrary given initial condition

$$x(\tau) = x_\tau, \qquad (\tau, x_\tau) \in R \times R^n.$$

In the metric of uniform convergence on R, the solution $x(t; \tau, x_\tau, F)$ depends continuously on the initial data (τ, x_τ) and on the right-hand side $F(t, x)$ as the parameter $\sigma = (\tau, x_\tau, F)$ ranges over the set

$$R \times R^n \times M \subset R \times R^n \times E_{\text{Lip}} = E_\sigma,$$

where M is an arbitrary fixed uniformly Lipschitzian subset of E_{Lip}, and the topology on $R \times R^n \times M$ is given by the seminorm $\|\cdot\|_w$ on E_σ.

The continuous dependence can be described in more detail as follows: Let M be an arbitrary uniformly Lipschitzian subset of E_{Lip}, and let

$$\hat{x}(t) = x(t; \hat{\tau}, \hat{x}_{\hat{\tau}}, \hat{F}), \qquad t \in R,$$

be the solution of the differential equation

$$\dot{x} = \hat{F}(t, x), \qquad \hat{F}(t, x) \in M,$$

with initial value

$$\hat{x}(\hat{\tau}) = \hat{x}_{\hat{\tau}}.$$

For every $\varepsilon > 0$, there exists a number $\delta > 0$ such that, for any point

$$(\tau, x_\tau, F(t, x)) \in R \times R^n \times M$$

that satisfies the condition

$$|\tau - \hat{\tau}| + |x_\tau - \hat{x}_{\hat{\tau}}| + \|F(t, x) - \hat{F}(t, x)\|_w \leq \delta,$$

the corresponding solution $x(t; \tau, x_\tau, F)$ satisfies the estimate

$$\|x(t; \tau, x_\tau, F) - x(t; \hat{\tau}, \hat{x}_{\hat{\tau}}, \hat{F})\| \leq \varepsilon.$$

We note that, in the statement of the continuous dependence, $F(t, x)$ ranges over a given, although arbitrary, uniformly Lipschitzian subset of E_{Lip}, and not over the entire space E_{Lip}.

We obtain the following result as a direct corollary of the theorem: If

$$\|F_1(t, x) - F_2(t, x)\|_w = 0,$$

then we have for all $(\tau, x_\tau) \in R \times R^n$

$$\|x(t; \tau, x_\tau, F_1) - x(t; \tau, x_\tau, F_2)\| = 0,$$

i.e., both solutions $x(t; \tau, x_\tau, F_1)$ and $x(t; \tau, x_\tau, F_2)$ coincide.

Proof. We reduce this theorem to Theorem 4.1 with the aid of a simple standard construction. Let C^n denote the linear space of all n-dimensional continuous functions on the t-axis which have finite limits as $t \to \pm\infty$. We introduce on C^n the norm

$$\|x(t)\| = \sup_{t \in R} |x(t)|, \qquad x(t) \in C^n,$$

which transforms C^n into a complete metric space with the metric

$$\rho(x(t), y(t)) = \|x(t) - y(t)\|.$$

We define a family of mappings $\varphi(x(\cdot); \sigma)$ of C^n into *itself* which depends on the parameter

$$\sigma = (\tau, x_\tau, F(t, x)) \in R \times R^n \times E_{\text{Lip}}$$

by the formula

$$x(\cdot) \to \varphi(x(\cdot); \sigma) = \varphi(x(\cdot); \tau, x_\tau, F) = x_\tau + \int_\tau^t F(\theta, x(\theta)) \, d\theta = y(t; x(\cdot), \sigma), \qquad t \in R.$$

Every fixed point of this mapping

$$x_\sigma(t) = x(t; \sigma) = x_\tau + \int_\tau^t F(\theta, x(\theta; \sigma)) \, d\theta, \qquad t \in R,$$

is the solution of the differential equation

$$\dot{x} = F(t, x)$$

Solutions of Differential Equations: Existence

with initial value $x(\tau; \sigma) = x_\tau$. Therefore, Theorem 4.2 will follow from Theorem 4.1 if we prove that, for any uniformly Lipschitzian set $M \subset E_{\text{Lip}}$, some iteration $\varphi^{(p)}(x(\cdot), \sigma)$ is a uniform contraction and is a continuous family of mappings of C^n into itself when the parameter σ ranges over the set

$$\Sigma = R \times R^n \times M \subset E_\sigma. \tag{4.5}$$

Here, according to our convention, the convergence $\sigma \to \hat{\sigma}$ means that

$$\|\sigma - \hat{\sigma}\|_w = |\tau - \hat{\tau}| + |x_\tau - \hat{x}_{\hat{\tau}}| + \|F(t, x) - \hat{F}(t, x)\|_w \to 0.$$

We shall now prove by induction that an arbitrary iteration $\varphi^{(p)}(x(\cdot), \sigma)$, $p = 1, 2, \ldots$ depends continuously on σ on any set (4.5) and that, if p is sufficiently large, then $\varphi^{(p)}(x(\cdot), \sigma)$ is a uniform contraction family on any set of the form (4.5).

In order to prove the continuity, we use the notation

$$\varphi(x(\cdot), \sigma) = y_1(t; x(\cdot), \sigma) = x_\tau + \int_\tau^t F(\theta, x(\theta))\, d\theta,$$

$$\varphi^{(p+1)}(x(\cdot), \sigma) = y_{p+1}(t; x(\cdot), \sigma) = x_\tau + \int_\tau^t F(\theta, y_p(\theta; x(\cdot), \sigma))\, d\theta, \quad p = 1, 2, \ldots,$$

and write the following obvious estimate:

$$|y_1(t; x(\cdot), \sigma) - y_1(t; x(\cdot), \hat{\sigma})| \leq |x_\tau - \hat{x}_{\hat{\tau}}| + \left|\int_\tau^{\hat{\tau}} |\hat{F}(\theta, x(\theta))|\, d\theta\right|$$

$$+ \left|\int_\tau^t \{F(\theta, x(\theta)) - \hat{F}(\theta, x(\theta))\}\, d\theta\right|.$$

The first two terms on the right-hand side obviously tend to zero as $\sigma \to \hat{\sigma}$. Therefore, in order to prove the continuous dependence of $\varphi(x(\cdot), \sigma)$ on $\sigma \in \Sigma$, it is sufficient to show that for $F(t, x)$ in the set M,

$$J = \max_{\theta_1, \theta_2 \in R} \left|\int_{\theta_1}^{\theta_2} \{F(\theta, x(\theta)) - \hat{F}(\theta, x(\theta))\}\, d\theta\right| \to 0 \tag{4.6}$$

as $\|F(t, x) - \hat{F}(t, x)\|_w \to 0$.

The function $x(t)$ has finite limits as $t \to \pm \infty$. Therefore for every $\varepsilon > 0$, there exists an interval $[t_1, t_2]$ such that, if we set $\hat{x}(t) = x(t)$ for t with

$t_1 \leq t \leq t_2$, $\hat{x}(t) = x(t_1)$ for $t \leq t_1$, and $\hat{x}(t) = x(t_2)$ for $t \geq t_2$, then we have

$$\max_{t \in R} |\hat{x}(t) - x(t)| = \|\hat{x}(t) - x(t)\| \leq \varepsilon.$$

Hence

$$J \leq \max_{\theta_1, \theta_2 \in R} \left| \int_{\theta_1}^{\theta_2} \{F(\theta, \hat{x}(\theta)) - \hat{F}(\theta, \hat{x}(\theta))\} \, d\theta \right| + \|\hat{x}(t) - x(t)\| \left\{ \int_R (L_F(t) + L_{\hat{F}}(t)) \, dt \right\}$$

$$\leq \max_{\theta_1, \theta_2 \in R} \left| \int_{\theta_1}^{\theta_2} \{F(\theta, \hat{x}(\theta)) - \hat{F}(\theta, \hat{x}(\theta))\} \, d\theta \right| + 2C\varepsilon,$$

where C is the constant of (4.4) for the uniformly Lipschitzian set under consideration. This constant does not depend on F or $\hat{F}$. The first term on the right-hand side tends to zero as $\|F(t, x) - \hat{F}(t, x)\|_w \to 0$ on the basis of Assertion 4.1. Since $\varepsilon > 0$ is arbitrary, we obtain that, if $F(t, x)$ always remains in M and if $\|F(t, x) - \hat{F}(t, x)\|_w \to 0$, then $J \to 0$. This proves the continuity of the family $\varphi(x(\cdot), \sigma)$ on any set of the form (4.5).

We shall now prove the continuity of the family $\varphi^{(p+1)}(x(\cdot), \sigma)$, assuming that the family $\varphi^{(p)}(x(\cdot), \sigma)$ is continuous. We write the following estimates:

$$|y_{p+1}(t; x(\cdot), \sigma) - y_{p+1}(t; x(\cdot), \hat{\sigma})|$$

$$\leq |x_\tau - \hat{x}_{\hat{\tau}}| + \left| \int_\tau^{\hat{\tau}} |\hat{F}(\theta, y_p(\theta; x(\cdot), \hat{\sigma}))| \, d\theta \right|$$

$$+ \left| \int_\tau^t |F(\theta, y_p(\theta; x(\cdot), \sigma)) - F(\theta, y_p(\theta; x(\cdot), \hat{\sigma}))| \, d\theta \right|$$

$$+ \left| \int_\tau^t \{F(\theta, y_p(\theta; x(\cdot), \hat{\sigma})) - \hat{F}(\theta, y_p(\theta; x(\cdot), \hat{\sigma}))\} \, d\theta \right|$$

$$\leq |x_\tau - \hat{x}_{\hat{\tau}}| + \left| \int_\tau^{\hat{\tau}} |\hat{F}(\theta, y_p(\theta; x(\cdot), \hat{\sigma}))| \, d\theta \right| + \|y_p(t; x(\cdot), \sigma) - y_p(t; x(\cdot), \hat{\sigma})\| \int_R L_F(t) \, dt$$

$$+ \left| \int_\tau^t \{F(\theta, y_p(\theta; x(\cdot), \hat{\sigma})) - \hat{F}(\theta, y_p(\theta; x(\cdot), \hat{\sigma}))\} \, d\theta \right|.$$

Solutions of Differential Equations: Existence

Obviously, the first two terms on the right-hand side of this inequality tend to zero as σ tends to $\hat{\sigma}$ through values in the set (4.5). The third term tends to zero by the induction assumption,

$$\|y_p(t; x(\cdot), \sigma) - y_p(t; x(\cdot), \hat{\sigma})\| \to 0,$$

and by virtue of the uniform boundedness of all the integrals

$$\int_R L_F(t)\, dt.$$

Finally, it can be proved in exactly the same way that relation (4.6) was proved, that for $\|F(t, x) - \hat{F}(t, x)\|_w \to 0$, $F(t, x) \in M$:

$$\max_{\theta_1 \theta_2 \in R} \left| \int_{\theta_1}^{\theta_2} \{F(\theta, y_p(\theta; x(\cdot), \hat{\sigma})) - \hat{F}(\theta, y_p(\theta; x(\cdot), \hat{\sigma}))\}\, d\theta \right| \to 0.$$

This concludes the proof of continuity.

We shall now prove that, for a sufficiently large p, $\varphi^{(p)}(x(\cdot), \sigma)$ is a uniform contraction family of mappings of the space C^n into itself for any set of parameters σ of the form (4.5). To this end, we estimate the difference

$$\|\varphi^{(p)}(x'(\cdot), \sigma) - \varphi^{(p)}(x''(\cdot), \sigma)\|$$
$$= \|y_p(t; x'(\cdot), \sigma) - y_p(t; x''(\cdot), \sigma)\|, \quad x'(t), x''(t) \in C^n, \sigma \in R \times R^n \times M.$$

We have

$$|y_p(t; x', \sigma) - y_p(t; x'', \sigma)| \leq \left| \int_\tau^t |F(\theta, y_{p-1}(\theta; x', \sigma)) - F(\theta, y_{p-1}(\theta; x'', \sigma))|\, d\theta \right|$$

$$\leq \left| \int_\tau^t L_F(\theta) |y_{p-1}(\theta; x', \sigma) - y_{p-1}(\theta; x'', \sigma)|\, d\theta \right|$$

$$= \left| \int_\tau^t L_F(\theta_1) |y'_{p-1}(\theta_1) - y''_{p-1}(\theta_1)|\, d\theta_1 \right|$$

$$\leq \left| \int_\tau^t L_F(\theta_1)\, d\theta_1 \int_\tau^{\theta_1} L_F(\theta_2) |y'_{p-2}(\theta_2) - y''_{p-2}(\theta_2)|\, d\theta_2 \right| \leq \cdots$$

$$\leq \left| \int_\tau^t L_F(\theta_1)\,d\theta_1 \int_\tau^{\theta_1} L_F(\theta_2)\,d\theta_2 \int_\tau^{\theta_2} \cdots \int_\tau^{\theta_{p-1}} L_F(\theta_p)|x'(\theta_p)-x''(\theta_p)|\,d\theta_p \right|$$

$$\leq \|x'(t)-x''(t)\| \left| \int_\tau^t L_F(\theta_1)\,d\theta_1 \int_\tau^{\theta_1} L_F(\theta_2)\,d\theta_2 \int_\tau^{\theta_2} \cdots \int_\tau^{\theta_{p-1}} L_F(\theta_p)\,d\theta_p \right|.$$

It is easy to prove by induction that

$$I_p(t,\tau) = \int_\tau^t L_F(\theta_1)\,d\theta_1 \int_\tau^{\theta_2} L_F(\theta_2)\,d\theta_2 \int_\tau^{\theta_2} \cdots \int_\tau^{\theta_{p-1}} L_F(\theta_p)\,d\theta_p$$

$$= \frac{1}{p!}\left(\int_\tau^t L_F(\theta)\,d\theta\right)^p.$$

Indeed, integration by parts yields

$$I_{p+1}(t,\tau) = \int_\tau^t L_F(\theta) I_p(\theta,\tau)\,d\theta$$

$$= \int_\tau^t L_F(\theta) \frac{1}{p!}\left\{\int_\tau^\theta L_F(\theta')\,d\theta'\right\}^p d\theta$$

$$= \frac{1}{p!}\left(\int_\tau^t L_F(\theta')\,d\theta'\right)^p \int_\tau^t L_F(\theta')\,d\theta'$$

$$-\frac{1}{p!}\int_\tau^t p\left(\int_\tau^\theta L_F(\theta')\,d\theta'\right)^{p-1} L_F(\theta)\left(\int_\tau^\theta L_F(\theta')\,d\theta'\right) d\theta$$

$$= \frac{1}{p!}\left(\int_\tau^t L_F(\theta')\,d\theta'\right)^{p+1} - \frac{1}{(p-1)!}\int_\tau^t L_F(\theta)\left(\int_\tau^\theta L_F(\theta')\,d\theta'\right)^p d\theta$$

$$= \frac{1}{p!}\left(\int_\tau^t L_F(\theta')\,d\theta'\right)^{p+1} - \frac{p!}{(p-1)!}\int_\tau^t L_F(\theta) I_p(\theta,\tau)\,d\theta$$

$$= \frac{1}{p!}\left(\int_\tau^t L_F(\theta)\,d\theta\right)^{p+1} - p I_{p+1}(t,\tau).$$

Hence

$$I_{p+1}(t,\tau) = \frac{1}{(p+1)!}\left(\int_\tau^t L_F(\theta)\,d\theta\right)^{p+1}.$$

The formula proved allows us to write the estimate

$$\|y_p(t; x', \sigma) - y_p(t; x'', \sigma)\| \leq \|x'(t) - x''(t)\| \frac{1}{p!}\left(\int_R L_F(\theta)\,d\theta\right)^p.$$

Since

$$\int_R L_F(\theta)\,d\theta \leq \text{const},$$

where the constant is independent of $F(t, x) \in M$, we obtain the final estimate

$$\rho(\varphi^{(p)}(x'(\cdot), \sigma), \varphi^{(p)}(x''(\cdot), \sigma)) \leq \frac{\text{const}^p}{p!}\rho(x'(\cdot), x''(\cdot))$$

$$\forall x'(t), x''(t) \in C^n \quad \text{and} \quad \forall \sigma = (\tau, x_\tau, F(t, x)) \in R \times R^n \times M.$$

This concludes the proof of the theorem.

4.4. The Spaces $E_{\text{Lip}}(G)$

Before formulating the existence and continuous dependence theorem in the general case, we must give several general definitions.

We denote by $E_{\text{Lip}}(G)$, where G is an arbitrary open subset of $R \times R^n$, the linear space of functions $F(t, x)$ that are defined on G and have the following property: If $\alpha(t, x)$ is a scalar-valued continuous function on $R \times R^n$ with compact support in G, then

$$\alpha(t, x)F(t, x) \in E_{\text{Lip}},$$

where the function $\alpha(t, x)F(t, x)$ is equal to zero outside of the support of the function $\alpha(t, x)$.

For every compact set $K \subset G$, we can define a seminorm $\|\cdot\|_{w,K}$ on $E_{\text{Lip}}(G)$, which is an analog of the seminorm $\|\cdot\|_w$ on E_{Lip}, in the following way: Let $\alpha_K(t, x)$ be a continuously differentiable function on $R \times R^n$ with values between 0 and 1,

$$0 \leq \alpha_K(t, x) \leq 1,$$

which takes on the value 1 on K and has a compact support in G. If we assume, as above, that the function $\alpha_K(t, x)F(t, x)$ with $F(t, x) \in E_{\text{Lip}}(G)$ is defined on the entire space $R \times R^n$—by setting it equal to zero outside of the support of $\alpha_K(t, x)$—then the seminorm $\|\alpha_K(t, x)F(t, x)\|_w$ is defined.

We set

$$\|F(t, x)\|_{w,K} = \inf_{\alpha_K} \|\alpha_K(t, x)F(t, x)\|_w,$$

where the infimum is taken over all the functions $\alpha_K(t, x)$ which have the indicated form. Obviously, if $K_1 \subset K_2 \subset G$, then

$$\|F(t, x)\|_{w,K_1} \leq \|F(t, x)\|_{w,K_2} \qquad \forall F(t, x) \in E_{\text{Lip}}(G).$$

If $K \subset G$ is the compact support of a function $F(t, x) \in E_{\text{Lip}}$, then the restriction of $F(t, x)$ to G belongs to $E_{\text{Lip}}(G)$ [it will be also denoted by $F(t, x)$], and

$$\|F(t, x)\|_{w,K} = \|F(t, x)\|_w.$$

Throughout the sequel, $\alpha_K(t, x)$ will denote an arbitrary, scalar-valued, continuously differentiable function which has a compact support in the set G under consideration and whose values are contained between 0 and 1 and equal to 1 on the compact set K. Only additional properties of such a function will be mentioned separately.

The following definition gives an analogue of the notion of a uniformly Lipschitzian set: A set of functions $M \subset E_{\text{Lip}}(G)$ is said to be *uniformly Lipschitzian on a neighborhood of the set* $K \subset G$ if there exists a compact neighborhood $V_K \subset G$ of the set K such that, for any scalar-valued continuously differentiable function $\alpha(t, x)$ with a support contained in V_K, the set

$$\alpha(t, x)M = \{\alpha(t, x)F(t, x) : F(t, x) \in M\} \subset E_{\text{Lip}}$$

is uniformly Lipschitzian in E_{Lip}.

A set $M \subset E_{\text{Lip}}(G)$ is said to be *uniformly Lipschitzian* in $E_{\text{Lip}}(G)$ if it is uniformly Lipschitzian on a neighborhood of any compact set $K \subset G$. If all the functions contained in $M \subset E_{\text{Lip}}(G)$ have compact supports in G, then we can assume that $M \subset E_{\text{Lip}}$. A set M can be uniformly Lipschitzian in E_{Lip} and, nevertheless, not be uniformly Lipschitzian in $E_{\text{Lip}}(G)$ without additional constraints imposed on M.

The following constraint, which may be called the *integral uniform boundedness* of a set M, guarantees that this set is uniformly Lipschitzian in $E_{\text{Lip}}(G)$ if it is uniformly Lipschitzian in E_{Lip}: A set $M \subset E_{\text{Lip}}$ will be called

Solutions of Differential Equations: Existence

integrally uniformly bounded if every function $F(t, x) \in M$ has a majorant $m_F(t)$

$$|F(t, x)| \leq m_F(t) \qquad \forall (t, x) \in R \times R^n$$

such that

$$\int_R m_F(t)\, dt \leq C,$$

where the constant C does not depend on $F(t, x) \in M$.

In order to show that a set M which is uniformly Lipschitzian in E_{Lip} and integrally uniformly bounded is also uniformly Lipschitzian in $E_{\text{Lip}}(G)$, we note that, for every scalar-valued continuously differentiable function $\alpha(t, x)$ with a compact support in G, there exists a constant D such that

$$|\alpha(t, x') - \alpha(t, x'')| \leq D|x' - x''| \qquad \forall (t, x'), (t, x'') \in G.$$

Therefore, we have the following estimates for every $F(t, x) \in M$:

$$|\alpha(t, x')F(t, x') - \alpha(t, x'')F(t, x'')|$$

$$\leq |\alpha(t, x')| \cdot |F(t, x') - F(t, x'')| + |\alpha(t, x') - \alpha(t, x'')| \cdot |F(t, x'')|$$

$$\leq \max_{(t,x)} |\alpha(t, x)| \cdot |x' - x''| \int_R L_F(t) + D|x' - x''| \int_R m_F(t).$$

These estimates prove the assertion, because the number

$$\max_{(t,x)} |\alpha(t, x)| \int_R L_F(t)\, dt + D \int_R m_F(t)\, dt$$

is no greater than a constant which does not depend on $F(t, x) \in M$.

A criterion for the property of being uniformly Lipschitzian in $E_{\text{Lip}}(G)$ which is of importance in control problems is presented in Chapter 6.

We shall identify an n-dimensional function $x(t)$ defined on an interval $[t_1, t_2]$ with the curve

$$\{(t, x(t)): t \in [t_1, t_2]\} \subset R \times R^n,$$

and we shall say, e.g., that this curve is *absolutely continuous* if the function $x(t)$ with $t_1 \leq t \leq t_2$ is. If

$$(t, x(t)) \in G \qquad \forall t \in [t_1, t_2],$$

then we shall say that the curve $x(t)$, $t_1 \leq t \leq t_2$, lies in G.

Any absolutely continuous curve $x(t)$, $t_1 \leq t \leq t_2$, which lies in G and satisfies the equality

$$\dot{x}(t) = F(t, x(t))$$

for almost all $t \in [t_1, t_2]$ is said to be a *solution*, or a *trajectory* of the differential equation

$$\dot{x} = F(t, x), \qquad F(t, x) \in E_{\text{Lip}}(G). \tag{4.7}$$

The solution of the integral equation

$$x(t) = x_\tau + \int_\tau^t F(\theta, x(\theta)) \, d\theta$$

is also defined in the usual way, and the equivalence of this equation with the differential equation (4.7) together with the *initial condition*

$$x(\tau) = x_\tau, \qquad (\tau, x_\tau) \in G,$$

for the solution is established.

If equation (4.7) and a solution $\tilde{x}(t)$ of this equation, $t_1 \leq t \leq t_2$, are given, then every equation of the form

$$\dot{x} = F(t, x) + \delta F(t, x), \qquad \delta F(t, x) \in E_{\text{Lip}}(G),$$

will be called a *perturbed equation*, and the function $\delta F(t, x)$ on the right-hand side will be called a perturbation. An arbitrary solution $x(t)$, $t_1 \leq t \leq t_2$, of the perturbed equation will be called a *perturbed solution*. The difference

$$\Delta x(t) = x(t) - \tilde{x}(t), \qquad t \in [t_1, t_2],$$

will be called a *perturbation* of the initial solution $\tilde{x}(t)$ on the time interval under consideration.

4.5. The Existence and Continuous Dependence Theorems for Solutions of Differential Equations in the General Case

Theorem 4.3. Let there be given the differential equation

$$\dot{x}(t) = F(t, x), \qquad F(t, x) \in E_{\text{Lip}}(G). \tag{4.8}$$

For every compact set $K \subset G$, there exists a positive number $\eta > 0$ such that, if the initial data point (τ, x_τ) belongs to K, then equation (4.8) has a unique solution $x(t)$ that is defined on the interval $\tau - \eta \leq t \leq \tau + \eta$ and satisfies the initial condition $x(\tau) = x_\tau$.

Solutions of Differential Equations: Existence

Proof. Let V_K be an arbitrary compact neighborhood of the set K which lies in G. We take a function $\alpha_{V_K}(t, x)$ and consider the differential equation

$$\dot{x} = \alpha_{V_K}(t, x) F(t, x). \tag{4.9}$$

An arbitrary solution of this equation with the initial data $(\tau, x_\tau) \in K$ will be denoted by

$$x(t) = x(t; \tau, x_\tau), \qquad t \in R.$$

Such a solution exists and is unique on the basis of Theorem 4.2.

Since $(\tau, x_\tau) \in K$, the function $\alpha_{V_K}(t, x)$ is equal to 1 on a neighborhood of the point (τ, x_τ). Therefore, there exists a number $\Delta > 0$ such that the function $x(t)$ satisfies the equality

$$\dot{x}(t) = F(t, x(t))$$

for almost all points of the interval $\tau - \Delta \leq t \leq \tau + \Delta$. Our aim is to show that a $\Delta > 0$ can be chosen the same for all points $(\tau, x_\tau) \in K$.

We denote by $m(t)$ a majorant of the function $\alpha_{V_K}(t, x) F(t, x)$,

$$|\alpha_{V_K}(t, x) F(t, x)| \leq m(t) \qquad \forall (t, x) \in R \times R^n.$$

Let d be the distance from K to the boundary of V_K. Let $\eta > 0$ be so small that $\eta < d/2$ and

$$\int_{t-\eta}^{t+\eta} m(\theta) \, d\theta \leq d/2 \qquad \forall t \in R.$$

For $|t - \tau| \leq \eta$, we obtain the estimate

$$|x(t; \tau, x_\tau) - x_\tau| = \left| \int_\tau^t \alpha_{V_K}(\theta, x(\theta; \tau, x_\tau)) F(\theta, x(\theta; \tau, x_\tau)) \, d\theta \right| \leq \int_{\tau-\eta}^{\tau+\eta} m(\theta) \, d\theta \leq \frac{d}{2}.$$

Since the distance from K to the boundary of V_K is d, we have

$$(t, x(t; \tau, x_\tau)) \in V_K \qquad \forall t \in [\tau - \eta, \tau + \eta]$$

independently of the choice of a point $(\tau, x_\tau) \in K$. Therefore, we obtain by virtue of the equality

$$\alpha_{V_K}(t, x) = 1 \qquad \forall (t, x) \in V_K$$

that, for $(\tau, x_\tau) \in K$, every curve

$$x(t) = x(t; \tau, x_\tau), \qquad \tau - \eta \leqslant t \leqslant \tau + \eta,$$

is a solution of the differential equation (4.8) with the initial condition $x(\tau) = x_\tau$. The uniqueness of this solution follows immediately from the uniqueness of the solution $x(t; \tau, x_\tau)$, $t \in R$, of equation (4.9).

Theorem 4.4. Let us be given the solution

$$\tilde{x}(t), \qquad t_1 \leqslant t \leqslant t_2,$$

of the differential equation

$$\dot{x} = F(t, x), \qquad F(t, x) \in E_{\text{Lip}}(G),$$

the initial data point $(\tilde{\tau}, \tilde{x}(\tilde{\tau})) = (\tilde{\tau}, \tilde{x}_{\tilde{\tau}})$ which lies on the solution, an arbitrary compact neighborhood

$$V_{\tilde{x}(t)} \subset G$$

of the curve $\tilde{x}(t)$, $t_1 \leqslant t \leqslant t_2$, and the set of perturbations

$$M \subset E_{\text{Lip}}(G),$$

which we assume to be uniformly Lipschitzian on the neighborhood $V_{\tilde{x}(t)}$. Then, for every $\varepsilon > 0$, there exists a $\delta > 0$ such that, if the initial data (τ, x_τ) and the perturbation $\delta F(t, x)$ satisfy the conditions

$$|\tau - \tilde{\tau}| + |x_\tau - \tilde{x}_{\tilde{\tau}}| + \|\delta F(t, x)\|_{w, V_{\tilde{x}(t)}} \leqslant \delta, \qquad \delta F(t, x) \in M,$$

then the perturbed equation

$$\dot{x} = F(t, x) + \delta F(t, x) \tag{4.10}$$

has the solution

$$x(t; \tau, x_\tau, \delta F), \qquad t_1 \leqslant t \leqslant t_2$$

[defined on the same interval as the initial solution $\tilde{x}(t)$] that satisfies the initial condition

$$x(\tau; \tau, x_\tau, \delta F) = x_\tau$$

and the estimate

$$\max_{t_1 \leqslant t \leqslant t_2} |x(t; \tau, x_\tau, \delta F) - \tilde{x}(t)| = \|x(t; \tau, x_\tau, \delta F) - \tilde{x}(t)\| \leqslant \varepsilon.$$

Solutions of Differential Equations: Existence

Proof. For convenience, we shall assume that $\delta F(t, x) = 0 \in M$. This can be done without loss of generality since, as it is easy to see, a set retains its property of being uniformly Lipschitzian when a finite number of functions are added to it.

Let a compact neighborhood $\tilde{V} \subset G$ of the curve $\tilde{x}(t)$, $t_1 \leq t \leq t_2$, be such that, for any scalar-valued continuously differentiable function $\alpha(t, x)$ whose support lies in $\tilde{V}$, the set

$$\alpha(t, x)M = \{\alpha(t, x)\delta F(t, x): \delta F(t, x) \in M\} \subset E_{\text{Lip}}$$

is uniformly Lipschitzian in E_{Lip}. We choose a compact neighborhood W of the curve $\tilde{x}(t)$ which lies in the intersection of the interiors of the neighborhoods $V_{\tilde{x}(t)}$ and $\tilde{V}$. Also, let a function $\alpha_W(t, x)$ be chosen so that its support is contained in the intersection of the neighborhoods $V_{\tilde{x}(t)}$ and $\tilde{V}$. Therefore, the set

$$\alpha_W(t, x)M = \{\alpha_W(t, x)\delta F(t, x): \delta F(t, x) \in M\} \subset E_{\text{Lip}} \tag{4.11}$$

is uniformly Lipschitzian in E_{Lip}.

We choose an arbitrary function $\alpha_{V_{\tilde{x}(t)}}(t, x)$. Obviously,

$$\alpha_W(t, x) = \alpha_{V_{\tilde{x}(t)}}(t, x)\alpha_W(t, x) \qquad \forall (t, x) \in R \times R^n.$$

We shall prove the inequality

$$\|\alpha_W(t, x)\delta F(t, x)\|_w \leq \text{const } \|\delta F(t, x)\|_{w, V_{\tilde{x}(t)}},$$

where the constant does not depend on the choice of $\delta F(t, x) \in M$.

We have

$$\int_{t'}^{t''} \alpha_W(t, x)\delta F(t, x)\, dt = \int_{t'}^{t''} \alpha_W(t, x)\alpha_{V_{\tilde{x}(t)}}(t, x)\delta F(t, x)\, dt$$

$$= \alpha_W(t'', x) \int_{t'}^{t''} \alpha_{V_{\tilde{x}(t)}}(t, x)\delta F(t, x)\, dt - \int_{t'}^{t''} \frac{\partial \alpha_W(t, x)}{\partial t} \left(\int_{t'}^{t} \alpha_{V_{\tilde{x}(t)}}(\theta, x)\delta F(\theta, x)\, d\theta \right) dt.$$

Therefore,

$$\left| \int_{t'}^{t''} \alpha_W(t, x)\delta F(t, x)\, dt \right| \leq \left| \int_{t'}^{t''} \alpha_{V_{\tilde{x}(t)}}(t, x)\delta F(t, x)\, dt \right|$$

$$+ \left| \int_{t'}^{t''} \left| \frac{\partial \alpha_W(t, x)}{\partial t} \right| dt \right| \max_{\substack{\theta', \theta'' \in R \\ x \in R^n}} \left| \int_{\theta'}^{\theta''} \alpha_{V_{\tilde{x}(t)}}(t, x)\delta F(t, x)\, dt \right|$$

$$\leqslant \left(1 + \int_R \left|\frac{\partial \alpha_W(t, x)}{\partial t}\right| dt\right) \|\alpha_{V_{\tilde{x}(t)}}(t, x)\delta F(t, x)\|_w$$

$$= \text{const } \|\alpha_{V_{\tilde{x}(t)}}(t, x)\delta F(t, x)\|_w.$$

Since the function $\alpha_{V_{\tilde{x}(t)}}(t, x)$ is arbitrary, we obtain the inequality

$$\left|\int_{t'}^{t''} \alpha_W(t, x)\delta F(t, x)\, dt\right| \leqslant \text{const } \|\delta F(t, x)\|_{w, V_{\tilde{x}(t)}},$$

which is equivalent to the inequality being proved.

We now consider the differential equation

$$\dot{x} = \alpha_W(t, x)F(t, x) + \alpha_W(t, x)\delta F(t, x), \qquad \delta F(t, x) \in M.$$

We denote its solution by

$$x(t) = x(t; \tau, x_\tau, \delta F), \qquad t \in R,$$

with initial value

$$x(\tau) = x_\tau.$$

The set (4.11) is uniformly Lipschitzian and contains the zero function. Therefore, on the basis of Theorem 4.2, for every $\varepsilon > 0$ there exists a $\Delta(\varepsilon) > 0$ such that we have

$$\|x(t; \tau, x_\tau, \delta F) - x(t; \tilde{\tau}, \tilde{x}_{\tilde{\tau}}, 0)\| \leqslant \varepsilon$$

for

$$|\tau - \tilde{\tau}| + |x_\tau - \tilde{x}_{\tilde{\tau}}| + \|\alpha_W(t, x)\delta F(t, x)\|_w \leqslant \Delta(\varepsilon), \qquad \delta F(t, x) \in M.$$

Since $\alpha_W(t, \tilde{x}(t)) = 1$ for all $t \in [t_1, t_2]$, it follows from the uniqueness of the solution that

$$x(t; \tilde{\tau}, \tilde{x}_{\tilde{\tau}}, 0) = \tilde{x}(t) \qquad \forall t \in [t_1, t_2].$$

Therefore, whenever $\varepsilon \leqslant d$, where d is the distance from the curve $\tilde{x}(t)$, $t_1 \leqslant t \leqslant t_2$, to the boundary of the neighborhood W, the following inclusion holds:

$$(t, x(t; \tau, x_{\tilde{\tau}}, \delta F)) \in W \qquad \forall t \in [t_1, t_2].$$

Solutions of Differential Equations: Existence 77

Since $\alpha_W(t, x) = 1 \ \forall (t, x) \in W$, the curve

$$x(t; \tau, x_\tau, \delta F), \qquad t_1 \leqslant t \leqslant t_2,$$

with $\varepsilon \leqslant d$ is the solution of the perturbed equation (4.10) with the initial value $x(\tau) = x_\tau$. Thus, for a given $\varepsilon > 0$, we can take for $\delta > 0$ the smaller of the numbers $\Delta(\varepsilon)$ and $\Delta(d)$.

5

The Variation Formula for Solutions of Differential Equations

5.1. The Spaces E_1 and $E_1(G)$

Let E_1 be the linear space of n-dimensional functions which are defined on $R \times R^n$ and satisfy the following conditions: Every function $F(t, x) \in E_1$ has a compact support and is continuously differentiable with respect to x for a fixed $t \in R$. For a fixed x, the (vector and matrix-valued) functions

$$F(t, x), \quad \frac{\partial F(t, x)}{\partial x} = F_x(t, x)$$

are measurable in t, and there exists a majorant $m_F(t)$ of $|F(t, x)| + |F_x(t, x)|$ which is summable on R,

$$|F(t, x)| + |F_x(t, x)| \leqslant m_F(t) \quad \forall (t, x) \in R \times R^n, \quad \int_R m_F(t)\, dt < \infty. \quad (5.1)$$

We shall show that every function in E_1 satisfies the Lipschitz condition (4.2), which will also prove the inclusion $E_1 \subset E_{\text{Lip}}$. We have

$$|F(t, x') - F(t, x'')| = \left| \int_0^1 \frac{d}{ds} F(t, x'' + s(x' - x''))\, ds \right|$$

$$\leqslant \int_0^1 |F_x(t, x'' + s(x' - x''))| \cdot |x' - x''|\, ds$$

$$\leqslant m_F(t) |x' - x''| \quad \forall (t, x'), (t, x'') \in R \times R^n.$$

We denote by $E_1(G)$, where G is an open subset of $R \times R^n$, the linear space of functions $F(t, x)$ which are defined on G and have the following property: If $\alpha(t, x)$ is a scalar-valued continuously differentiable function on $R \times R^n$ with compact support in G, then $\alpha(t, x)F(t, x) \in E_1$.

The following obvious inclusion holds:

$$E_1(G) \subset E_{\text{Lip}}(G).$$

Therefore, all the notions and facts related to E_{Lip} and $E_{\text{Lip}}(G)$ are transferred to E_1 and $E_1(G)$ in an automatic way.

We define a seminorm $\|\cdot\|_1$ on E_1 by the formula

$$\|F(t, x)\|_1 = \int_R \max_{x \in R^n} (|F(t, x)| + |F_x(t, x)|) \, dt.$$

In order to prove that the integrand is integrable in t, so that the definition is correct, we note that the continuity of the function $|F(t, x)| + |F_x(t, x)|$ implies that, if $x_1, x_2, \ldots$ is a sequence dense in R^n, then

$$\max_{x \in R^n} (|F(t, x)| + |F_x(t, x)|) = \sup_i (|F(t, x_i)| + |F_x(t, x_i)|).$$

Since on the right-hand side under the sup sign we have a countable sequence of functions measurable in t, the left-hand side of the equality is also measurable in t. Now the integrability follows from the existence of the majorant (5.1).

In addition to $\|\cdot\|_1$, the seminorm $\|\cdot\|_w$, which was considered in Chapter 4, is also defined on E_1. Obviously,

$$\|F(t, x)\|_w \leq \|F(t, x)\|_1 \qquad \forall F(t, x) \in E_1.^*$$

We define the seminorm $\|\cdot\|_{1,K}$ on $E_1(G)$ for an arbitrary compact set $K \subset G$ by the formula

$$\|F(t, x)\|_{1,K} = \inf \|\alpha_K(t, x) F(t, x)\|_1,$$

where the infimum is taken over all functions $\alpha_K(t, x)$ (see Chapter 4 for the notation). The following obvious inequality holds:

$$|F(t, x)\|_{w,K} \leq \|F(t, x)\|_{1,K} \qquad \forall F(t, x) \in E_1(G) \quad \text{and} \quad \forall K \subset G.$$

---

*As in the case of the seminorm $\|\cdot\|_w$, the seminorm $\|\cdot\|_1$ becomes a norm if we identify in E_1 any two functions whose difference is zero for every fixed x and for almost all $t \in R$. This is so because, as it is not difficult to show, the statement $\{\forall x \in R^n \; F(t, x) = 0 \text{ for almost all } t\}$ is equivalent to the equality $\|F(t, x)\|_1 = 0$.

Solutions of Differential Equations: Variation Formula

If we consider the differential equation

$$\dot{x} = F(t, x), \qquad F(t, x) \in E_1(G),$$

then, naturally, all perturbations $\delta F(t, x)$ on the right-hand side are taken from $E_1(G)$.

We shall now present a simple criterion for a subset of $E_1(G)$ to be uniformly Lipschitzian on a neighborhood of a given compact set $K \subset G$.

Assertion 5.1. Let K be a compact subset of G, and let $V_K \subset G$ be an arbitrary compact neighborhood of K. Then the set

$$M = \{F(t, x): \|F(t, x)\|_{1, V_K} \leq \text{const}\} \subset E_1(G),$$

where const denotes a fixed constant, is uniformly Lipschitzian on the neighborhood of K.

Proof. We take a scalar-valued continuously differentiable function $\alpha(t, x)$ with support in V_K, and we shall show that the set

$$\{\alpha(t, x)F(t, x): F(t, x) \in M\} \subset E_1 \subset E_{\text{Lip}}$$

is uniformly Lipschitzian in E_{Lip}.

We have for an arbitrary function $\alpha_{V_K}(t, x)$,

$$\alpha(t, x)F(t, x) = \alpha(t, x)\alpha_{V_K}(t, x)F(t, x) = \alpha(t, x)F_{V_K}(t, x),$$

where for brevity we used the notation

$$F_{V_K}(t, x) = \alpha_{V_K}(t, x)F(t, x).$$

We write the estimates

$$|\alpha(t, x')F(t, x') - \alpha(t, x'')F(t, x'')|$$

$$= \left| \int_0^1 \frac{d}{ds} \alpha(t, x'' + s(x' - x''))F_{V_K}(t, x'' + s(x' - x''))\, ds \right|$$

$$\leq \int_0^1 \left| \frac{\partial}{\partial x} \left\{ \alpha(t, x'' + s(x' - x''))F_{V_K}(t, x'' + s(x' - x'')) \right\} \right| \cdot |x' - x''|\, ds$$

$$\leq \left\{ \max_{x \in R^n} \left| \frac{\partial}{\partial x} \alpha(t, x)F_{V_K}(t, x) \right| \right\} |x' - x''| = L(t)|x' - x''|.$$

Now,

$$L(t) \leqslant \left\{ \max_{x \in R^n} |\alpha(t,x)| + \max_{x \in R^n} \left|\frac{\partial}{\partial x} \alpha(t,x)\right| \right\} \max_{x \in R^n} \left\{ |F_{V_K}(t,x)| + \left|\frac{\partial}{\partial x} F_{V_K}(t,x)\right| \right\}$$

$$\leqslant C_1 \max_{x \in R^n} \left\{ |F_{V_K}(t,x)| + \left|\frac{\partial}{\partial x} F_{V_K}(t,x)\right| \right\}.$$

Hence we obtain the inequality

$$\int_R L(t)\, dt \leqslant C_1 \int_R \max_{x \in R^n} \left\{ |F_{V_K}(t,x)| + \left|\frac{\partial}{\partial x} F_{V_K}(t,x)\right| \right\} dt.$$

Since the function $\alpha_{V_K}(t,x)$ is arbitrary, this implies the final estimate

$$\int_R L(t)\, dt \leqslant C_1 \cdot \text{const},$$

which completes the proof.

5.2. The Equation of Variation and the Variation Formula for the Solution

We consider the differential equation

$$\dot{x} = F(t,x), \qquad F(t,x) \in E_1(G). \tag{5.2}$$

Let $\tilde{x}(t)$, $t_1 \leqslant t \leqslant t_2$, be a solution of this equation, and let $V_{\tilde{x}(t)} \subset G$ be a compact neighborhood of the curve $\tilde{x}(t)$. According to the assertion just proved, the set of perturbations

$$\{\delta F(t,x):\ \|\delta F(t,x)\|_{1,V_{\tilde{x}(t)}} \leqslant 1\} \subset E_1(G)$$

is uniformly Lipschitzian on a neighborhood of the curve $\tilde{x}(t)$. Thus, if the initial value vector x_1 and the perturbation $\delta F(t,x)$ satisfy the inequality

$$|x_1 - \tilde{x}(t_1)| + \|\delta F(t,x)\|_{1,V_{\tilde{x}(t)}} \leqslant \varepsilon$$

and therefore, the inequality

$$|x_1 - \tilde{x}(t_1)| + \|\delta F(t,x)\|_{w,V_{\tilde{x}(t)}} \leqslant \varepsilon,$$

then, according to the continuous dependence theorem (Theorem 4.4), the

Solutions of Differential Equations: Variation Formula

perturbed equation

$$\dot{x} = F(t, x) + \delta F(t, x)$$

has, for all $\varepsilon \geq 0$ sufficiently small, the solution

$$x(t) = x(t; x_1, \delta F), \qquad t_1 \leq t \leq t_2,$$

with the initial value

$$x(t_1) = x_1.$$

Here we have, for $\varepsilon \to 0$,

$$\max_{t_1 \leq t \leq t_2} |x(t; x_1, \delta F) - \tilde{x}(t)| = \|x(t; x_1, \delta F) - \tilde{x}(t)\| \to 0.$$

Thus, if the number

$$|x_1 - \tilde{x}(t_1)| + \|\delta F(t, x)\|_{1, V_{\tilde{x}(t)}}$$

is small, then the differential equation

$$\Delta \dot{x} = F(t, \tilde{x}(t) + \Delta x) - F(t, \tilde{x}(t)) + \delta F(t, \tilde{x}(t) + \Delta x) \tag{5.3}$$

for the perturbation

$$\Delta x(t) = \Delta x(t; \delta x_1, \delta F) = x(t; x_1, \delta F) - \tilde{x}(t), \qquad \delta x_1 = x_1 - \tilde{x}(t_1),$$

has a solution on the interval $t_1 \leq t \leq t_2$ with the initial value

$$\Delta x(t_1) = \delta x_1.$$

Moreover,

$$\max_{t_1 \leq t \leq t_2} |\Delta x(t; \delta x_1, \delta F)| = \|\Delta x(t; \delta x_1, \delta F)\| \to 0$$

as $|\delta x_1| + \|\delta F(t, x)\|_{1, V_{\tilde{x}(t)}} \to 0$. This is all that can be said about the perturbation $\Delta x(t)$, if one only takes into account the condition

$$|x_1 - \tilde{x}(t_1)| + \|\delta F(t, x)\|_{w, V_{\tilde{x}(t)}} \to 0.$$

However, in the case under consideration, i.e., when

$$|x_1 - \tilde{x}(t_1)| + \|\delta F(t, x)\|_{1, V_{\tilde{x}(t)}} \to 0,$$

a much stronger assertion is true. We shall prove that $\Delta x(t)$ can be represented in the form

$$\Delta x(t; \delta x_1, \delta F) = \delta x(t; \delta x_1, \delta F) + \Delta_2 x(t; \delta x_1, \delta F), \qquad t_1 \leq t \leq t_2, \tag{5.4}$$

where $\delta x(t; \delta x_1, \delta F)$ satisfies the linear differential equation

$$\delta \dot{x} = F_x(t, \tilde{x}(t))\delta x + \delta F(t, \tilde{x}(t)) \qquad (5.5)$$

with the initial condition

$$\delta x(t_1; \delta x_1, \delta F) = \delta x_1,$$

and $\Delta_2 x(t; \delta x_1, \delta F)$ satisfies the condition

$$\max_{t_1 < t < t_2} |\Delta_2 x(t; \delta x_1, \delta F)| = \|\Delta_2 x(t; \delta x_1, \delta F)\| = o(|\delta x_1| + \|\delta F(t, x)\|_{1, V_{\tilde{x}(t)}}) \qquad (5.6)$$

where

$$\frac{o(\varepsilon)}{\varepsilon} \to 0 \qquad (\varepsilon \to 0).$$

The function $\delta x(t; \delta x_1, \delta F)$, as a solution of the linear equation (5.5), depends linearly on $(\delta x_1, \delta F)$. Therefore, on the basis of (5.6), this function is the "main linear part," or the *differential*, of the mapping

$$(\delta x_1, \delta F) \to \Delta x(t; \delta x_1, \delta F), \qquad t_1 \leq t \leq t_2,$$

at the point $(\delta x_1, \delta F) = 0$, if the seminorm

$$\|(\delta x_1, \delta F)\| = |\delta x_1| + \|\delta F(t, x)\|_{1, V_{\tilde{x}(t)}} \qquad (5.7)$$

is introduced on the linear space of all pairs $(\delta x_1, \delta F)$, and the norm of uniform convergence on the interval $[t_1, t_2]$ is introduced on the space of functions $\Delta x(t; \delta x_1, \delta F)$.

The function $\delta x(t; \delta x_1, \delta F)$, $t_1 \leq t \leq t_2$, is said to be the *variation* of the solution $\tilde{x}(t)$ when the initial value $\tilde{x}(t_1)$ is perturbed (varied) by δx_1, and the right-hand side $F(t, x)$ of equation (5.2) is perturbed (varied) by $\delta F(t, x)$.

The linear nonhomogeneous equation (5.5) is called the *equation of variation* for (5.2) *along the trajectory* $\tilde{x}(t)$, $t_1 \leq t \leq t_2$, which corresponds to the perturbation $\delta F(t, x)$. This equation is obtained as a result of the "linearization" of the nonlinear equation (5.3) for the perturbation $\Delta x(t)$.

The homogeneous equation

$$\delta \dot{x} = F_x(t, \tilde{x}(t))\delta x \qquad (5.8)$$

is called the *homogeneous equation of variation for* (5.2) *along* $\tilde{x}(t)$, $t_1 \leq t \leq t_2$.

All that we have said thus far can be expressed in the following basic theorem of this chapter:

Solutions of Differential Equations: Variation Formula

Theorem 5.1. For all sufficiently small values of (5.7), equation (5.3) for the perturbation $\Delta x(t)$ has the solution

$$\Delta x(t) = \Delta x(t; \delta x_1, \delta F), \qquad t_1 \leq t \leq t_2,$$

with the initial value

$$\Delta x(t_1) = \delta x_1.$$

This solution can be represented in the form (5.4). Here, the variation $\delta x(t; \delta x_1, \delta F)$ satisfies the equation of variation (5.5) with the initial condition

$$\delta x(t_1; \delta x_1, \delta F) = \delta x_1,$$

and the function $\Delta_2 x(t; \delta x_1, \delta F)$, $t_1 \leq t \leq t_2$, satisfies the condition (5.6).

Before we begin the proof of the theorem, let us make several additional remarks. We denote by $\Gamma(t)$ the fundamental matrix of solutions of equation (5.8) that reduces to the identity matrix at $t = t_1$. We denote by $G(t)$ the matrix inverse to $\Gamma(t)$ (see Section 5.5 for the construction of fundamental matrices). We have

$$\dot{\Gamma}(t) = F_x(t, \tilde{x}(t))\Gamma(t), \qquad \dot{G}(t) = -G(t)F_x(t, \tilde{x}(t)),^* \qquad t_1 \leq t \leq t_2.$$

As a solution of equation (5.5), the variation $\delta x(t; \delta x_1, \delta F)$ can be expressed by the formula

$$\delta x(t; \delta x_1, \delta F) = \Gamma(t)\left(\delta x_1 + \int_{t_1}^{t} G(\theta)\delta F(\theta, \tilde{x}(\theta))\, d\theta \right), \qquad (5.9)$$

which is called the *formula for the variation of the solution* $\tilde{x}(t)$, $t_1 \leq t \leq t_2$. If we set $\delta F(t, x) = 0$ and vary the solution $\tilde{x}(t)$ by varying only its initial value by δx_1, i.e., if we are solving the homogeneous equation of variation, then we obtain

$$\delta x(t; \delta x_1, 0) = \Gamma(t)\delta x_1.$$

Assume that Π_{t_1} is a hyperplane in R^n which passes through the point $\tilde{x}(t_1)$. We represent Π_{t_1} in the form of displacements from the point $\tilde{x}(t_1)$ by an arbitrary vector $\delta x \in \Pi$, where Π is the corresponding $(n-1)$-dimensional

---

*In order to derive the second equation, we differentiate the identity matrix $G(t)\Gamma(t)$:

$$\dot{G}(t)\Gamma(t) + G(t)\dot{\Gamma}(t) = \{\dot{G}(t) + G(t)F_x(t, \tilde{x}(t))\}\Gamma(t) = 0.$$

Dividing by the nonsingular matrix $\Gamma(t)$, we obtain

$$\dot{G}(t) + G(t)F_x(t, \tilde{x}(t)) = 0.$$

subspace of R^n:

$$\Pi_{t_1} = \{\tilde{x}(t_1) + \delta x_1 : \delta x_1 \in \Pi\}.$$

The family of hyperplanes

$$\Pi_t = \{\tilde{x}(t) + \Gamma(t)\delta x_1 = \tilde{x}(t) + \delta x(t; \delta x_1, 0) : \delta x_1 \in \Pi\}, \qquad t_1 \leq t \leq t_2,$$

is called the translation of the hyperplane Π_{t_1} along $\tilde{x}(t)$, $t_1 \leq t \leq t_2$, with the aid of the homogeneous equation in variations (5.8). This expression gives, in fact, an entire family of hyperplanes Π_t, since the matrix $\Gamma(t)$ is nonsingular. Each hyperplane Π_t is obtained from the translation with the aid of the correspondence

$$\tilde{x}(t_1) + \delta x_1 \to \tilde{x}(t) + \Gamma(t)\delta x_1 \qquad \forall \delta x_1 \in \Pi, \qquad t_1 \leq t \leq t_2.$$

If a vector ξ is orthogonal to Π, then the vector

$$\psi(t) = \xi G(t), \qquad t_1 \leq t \leq t_2,$$

is orthogonal to the subspace $\Pi_t - \tilde{x}(t)$ because

$$\psi(t)\Gamma(t)\delta x_1 = \xi G(t)\Gamma(t)\delta x_1 = \xi \delta x_1 = 0.$$

Obviously, the function $\psi(t)$, $t_1 \leq t \leq t_2$, satisfies the differential equation

$$\dot{\psi} = -\psi F_x(t, \tilde{x}(t))$$

and the initial condition

$$\psi(t_1) = \xi.$$

The differential equation for $\psi(t)$ is said to be adjoint to the homogeneous equation of variation (5.8). The former equation is characterized by the fact that the scalar product of any solution $\chi(t)$ of this equation with any solution $\delta x(t)$ of equation (5.8) is constant,

$$\chi(t)\delta x(t) = \text{const}$$

(where, of course, the constant can depend on a solution). Indeed, the equality

$$\dot{\chi}(t)\delta x(t) + \chi(t)\delta \dot{x}(t) = \{\dot{\chi}(t) + \chi(t)F_x(t, \tilde{x}(t))\}\delta x(t) = 0$$

is equivalent to the equality

$$\dot{\chi}(t) = -\chi(t)F_x(t, \tilde{x}(t)),$$

since the vector δx_1 and, therefore, the vector $\delta x(t) = \Gamma(t)\delta x_1$, $t_1 \leq t \leq t_2$, are arbitrary.

5.3. Proof of Theorem 5.1

First, we shall prove that, for all values $|\delta x_1| + \|\delta F(t, x)\|_{1, V_{\tilde{x}(t)}}$ sufficiently small, the perturbation

$$\Delta x(t) = \Delta x(t; \delta x_1, \delta F), \quad t_1 \leq t \leq t_2,$$

satisfies the inequality

$$\|\Delta x(t)\| \leq C(|\delta x_1| + \|\delta F(t, x)\|_{1, V_{\tilde{x}(t)}}).$$

We rewrite equation (5.3) for $\Delta x(t)$ in the form

$$\Delta \dot{x}(t) = F(t, \tilde{x}(t) + \Delta x(t)) - F(t, \tilde{x}(t)) + \delta F(t, \tilde{x}(t) + \Delta x(t))$$

$$= F_x(t, \tilde{x}(t))\Delta x(t) + \delta F(t, \tilde{x}(t)) + \int_0^1 \delta F_x(t, \tilde{x}(t) + s\Delta x(t))\Delta x(t)\, ds$$

$$+ \int_0^1 \{F_x(t, \tilde{x}(t) + s\Delta x(t)) - F_x(t, \tilde{x}(t))\}\Delta x(t)\, ds$$

$$= F_x(t, \tilde{x}(t))\Delta x(t) + \delta F(t, \tilde{x}(t)) + R_1(t) + R_2(t),$$

where

$$R_1(t) = \int_0^1 \delta F_x(t, \tilde{x}(t) + s\Delta x(t))\Delta x(t)\, ds,$$

$$R_2(t) = \int_0^1 \{F_x(t, \tilde{x}(t) + s\Delta x(t)) - F_x(t, \tilde{x}(t))\}\Delta x(t)\, ds.$$

The equation thus obtained can be viewed as a linear nonhomogeneous equation in $\Delta x(t)$ with the nonhomogeneous term

$$\delta F(t, \tilde{x}(t)) + R_1(t) + R_2(t).$$

Taking into account the initial value $\Delta x(t_1) = \delta x_1$, we can therefore write

$$\Delta x(t) = \Gamma(t)\left\{\delta x_1 + \int_{t_1}^t G(\theta)[\delta F(\theta, \tilde{x}(\theta)) + R_1(\theta) + R_2(\theta)]\, d\theta\right\}. \quad (5.10)$$

Let us estimate the absolute value of the expression

$$\Gamma(t) \int_{t_1}^t G(\theta)[\delta F(\theta, \tilde{x}(\theta)) + R_1(\theta) + R_2(\theta)]\, d\theta.$$

Let
$$|\Gamma(t)| \leq C_1, \quad |G(t)| \leq C_1, \quad t_1 \leq t \leq t_2.$$

Then
$$\left|\Gamma(t) \int_{t_1}^{t} G(\theta)\delta F(\theta, \tilde{x}(\theta))\, d\theta\right| \leq C_1^2 \int_{t_1}^{t_2} \max_{x \in V_{\tilde{x}(t)}} |\delta F(\theta, \tilde{x}(\theta))|\, d\theta \leq C_1^2 \|\delta F(t, x)\|_{1, V_{\tilde{x}(t)}}. \tag{5.11}$$

For $\|\Delta x(t)\|$ sufficiently small,
$$\tilde{x}(t) + s\Delta x(t) \in V_{\tilde{x}(t)} \quad \forall (s, t) \in [0, 1] \times [t_1, t_2].$$

Therefore,
$$\left|\Gamma(t) \int_{t_1}^{t} G(\theta) R_1(\theta)\, d\theta\right| \leq C_1^2 \|\Delta x(t)\| \int_{t_1}^{t_2} \max_{x \in V_{\tilde{x}(t)}} |\delta F_x(t, x)|\, dt$$
$$\leq C_1^2 \|\delta F(t, x)\|_{1, V_{\tilde{x}(t)}} \|\Delta x(t)\|. \tag{5.12}$$

Furthermore,
$$\left|\Gamma(t) \int_{t_1}^{t} G(\theta) R_2(\theta)\, d\theta\right| \leq C_1^2 \|\Delta x(t)\| \int_{t_1}^{t_2} dt \int_0^1 |F_x(t, \tilde{x}(t) + s\Delta x(t)) - F_x(t, \tilde{x}(t))|\, ds$$
$$= C_1^2 \|\Delta x(t)\| \vartheta(\Delta x(t)), \tag{5.13}$$

where $\vartheta(\Delta x(t))$ is a functional of $\Delta x(t)$, $t_1 \leq t \leq t_2$,
$$\vartheta(\Delta x(t)) = \int_{t_1}^{t_2} dt \int_0^1 |F_x(t, \tilde{x}(t) + s\Delta x(t)) - F_x(t, \tilde{x}(t))|\, ds.$$

We shall prove that $\vartheta(\Delta x(t))$ tends to zero as $\|\Delta x(t)\| \to 0$. We represent ϑ in the form of the following double integral:
$$\vartheta(\Delta x(t)) = \int_0^1 \int_{t_1}^{t_2} |F_x(t, \tilde{x}(t) + s\Delta x(t)) - F_x(t, \tilde{x}(t))|\, dt\, ds.$$

For a fixed $(s, t) \in [0, 1] \times [t_1, t_2]$ and for $\|\Delta x\| \to 0$, the integrand tends to zero and is majorized by the following summable function:
$$|F_x(t, \tilde{x}(t) + s\Delta x(t)) - F_x(t, \tilde{x}(t))| \leq \alpha_{V_{\tilde{x}(t)}}(t, x)\{|F_x(t, \tilde{x}(t) + s\Delta x(t))| + |F_x(t, \tilde{x}(t))|\}$$
$$\leq 2m(t).$$

Here, we can use for $m(t)$ a majorant of the function

$$\left| \alpha_{V_{\tilde{x}(t)}}(t, x) F(t, x) \right| + \left| \frac{\partial}{\partial x} \alpha_{V_{\tilde{x}(t)}}(t, x) F(t, x) \right|.$$

Therefore, according to the Lebesgue theorem on passing to the limit under the integral, we obtain

$$\lim_{\|\Delta x(t)\| \to 0} \vartheta(\Delta x(t)) = \int_0^1 \int_{t_1}^{t_2} \lim_{\|\Delta x\| \to 0} \left| F(t, \tilde{x}(t) + s\Delta x(t)) - F_x(t, \tilde{x}(t)) \right| dt \, ds = 0.$$

Thus, on the basis of (5.10) and the estimates (5.11)–(5.13) we can write

$$\|\Delta x(t)\| \leq C_1 |\delta x_1| + C_1^2 \|\delta F(t, x)\|_{1, V_{\tilde{x}(t)}} + \|\Delta x(t)\| \{ C_1^2 \vartheta(\Delta x(t))$$
$$+ C_1^2 \|\delta F(t, x)\|_{1, V_{\tilde{x}(t)}} \}.$$

Since $\vartheta(\Delta x(t)) \to 0$ as $|\delta x_1| + \|\delta F(t, x)\|_{1, V_{\tilde{x}(t)}} \to 0$, we have

$$\|\Delta x(t)\| \leq \frac{(1 + C_1)^2}{1 - C_1^2 \vartheta(\Delta x(t)) - C_1^2 \|\delta F(t, x)\|_{1, V_{\tilde{x}(t)}}} (|\delta x_1| + \|\delta F(t, x)\|_{1, V_{\tilde{x}(t)}})$$
$$\leq 2(1 + C_1)^2 (|\delta x_1| + \|\delta F(t, x)\|_{1, V_{\tilde{x}(t)}}). \tag{5.14}$$

This basic estimate expresses the fact that the order of $\|\Delta x(t)\|$ is no less than the first power of $|\delta x_1| + \|\delta F(t, x)\|_{1, V_{\tilde{x}(t)}}$ as $|\delta x_1| + \|\delta F(t, x)\|_{1, V_{\tilde{x}(t)}} \to 0$.

After the estimate (5.14) has been obtained and the relation $\vartheta(\Delta x(t)) \to 0$ proved, it is easy to obtain the representation (5.4) for $\Delta x(t)$. To this end, we define the function $\delta x(t; \delta x_1, \delta F)$ in (5.4) in accordance with the formula (5.9), i.e., as the solution of the equation of variation (5.5) with the initial value $\delta x(t_1) = \delta x_1$. We shall prove that in this case the difference

$$\Delta_2 x(t) = \Delta_2 x(t; \delta x_1, \delta F) = \Delta x(t; \delta x_1, \delta F) - \delta x(t; \delta x_1, \delta F), \qquad t_1 \leq t \leq t_2,$$

satisfies the condition

$$\frac{\|\Delta_2 x(t)\|}{|\delta x_1| + \|\delta F(t, x)\|_{1, V_{\tilde{x}(t)}}} \to 0, \qquad |\delta x_1| + \|\delta F(t, x)\|_{1, V_{\tilde{x}(t)}} \to 0.$$

We have

$$\Delta \dot{x} = F_x(t, \tilde{x}(t)) \Delta x(t) + \delta F(t, \tilde{x}(t)) + R_1(t) + R_2(t)$$

and

$$\delta \dot{x} = F_x(t, \tilde{x}(t)) \delta x(t) + \delta F(t, \tilde{x}(t)).$$

If we subtract the second equation from the first and use the definition

$\Delta_2 x(t) = \Delta x(t) - \delta x(t)$, we get

$$\Delta_2 \dot{x}(t) = F_x(t, \tilde{x}(t))\Delta_2 x(t) + R_1(t) + R_2(t).$$

The initial value $\Delta_2 x(t_1) = 0$, because $\delta x(t_1) = \delta x_1 = \Delta x(t_1)$. Therefore,

$$\Delta_2 x(t) = \Gamma(t) \int_{t_1}^{t} G(\theta)(R_1(\theta) + R_2(\theta))\, d\theta.$$

Making use of the estimates (5.12)–(5.14), we can write the inequalities

$$\|\Delta_2 x(t)\| \leq \{C_1^2 \vartheta(\Delta x(t)) + C_1^2 \|\delta F(t, x)\|_{1, V_{\tilde{x}(t)}}\} \|\Delta x(t)\|$$
$$\leq 2C_1^2(1 + C_1)^2 \{\vartheta(\Delta x(t)) + \|\delta F(t, x)\|_{1, V_{\tilde{x}(t)}}\}(|\delta x_1| + \|\delta F(t, x)\|_{1, V_{\tilde{x}(t)}}),$$

which complete the proof of Theorem 5.1.

5.4. A Counterexample

Let a family of perturbations $\delta F(t, x; \varepsilon)$, $0 \leq \varepsilon \leq 1$, be twice continuously differentiable with respect to x, uniformly Lipschitzian on a neighborhood of a solution $\tilde{x}(t)$, $t_1 \leq t \leq t_2$, of equation (5.2), and let it satisfy the condition

$$\|\delta F(t, x; \varepsilon)\|_{w, V_{\tilde{x}(t)}} + \|\delta F_x(t, x; \varepsilon)\|_{w, V_{\tilde{x}(t)}} \leq \varepsilon \cdot \text{const}. \tag{5.15}$$

Obviously, this condition is weaker than the condition

$$\|\delta F(t, x; \varepsilon)\|_{1, V_{\tilde{x}(t)}} \leq \varepsilon \cdot \text{const}. \tag{5.16}$$

Condition (5.15) asserts that the order of the seminorm on the left-hand side is no less than ε as $\varepsilon \to 0$. According to the continuous dependence theorem (Theorem 4.4), the perturbed equation

$$\dot{x} = F(t, x) + \delta F(t, x; \varepsilon)$$

has the following solution for all ε sufficiently small:

$$x(t; \varepsilon), \quad t_1 \leq t \leq t_2,$$
$$x(t_1; \varepsilon) \equiv \tilde{x}(t_1).$$

This solution tends to $\tilde{x}(t)$ as $\varepsilon \to 0$ uniformly on the interval $[t_1, t_2]$, i.e.,

$$\max_{t_1 \leq t \leq t_2} |x(t; \varepsilon) - \tilde{x}(t)| = \|x(t; \varepsilon) - \tilde{x}(t)\| = \|\Delta x(t; \varepsilon)\| \to 0.$$

The order of the perturbation $\Delta x(t; \varepsilon)$ is no less than ε. Indeed, expanding the

Solutions of Differential Equations: Variation Formula

right-hand side of the equation

$$\Delta \dot{x}(t) = F(t, \tilde{x} + \Delta x(t)) - F(t, \tilde{x}(t)) + \delta F(t, \tilde{x}(t) + \Delta x(t); \varepsilon)$$

in series in powers of Δx, we obtain

$$\Delta \dot{x} = F_x(t, \tilde{x}(t))\Delta x + \delta F(t, \tilde{x}(t); \varepsilon) + \delta F_x(t, \tilde{x}(t); \varepsilon)\Delta x + \cdots. \qquad (5.17)$$

Considering $\Delta x(t)$ as a solution of a linear equation with the nonhomogeneous term

$$\delta F(t, \tilde{x}(t); \varepsilon) + \delta F_x(t, \tilde{x}(t); \varepsilon)\Delta x + \cdots,$$

we can write

$$\Delta x(t; \varepsilon) = \Gamma(t) \int_{t_1}^{t} G(\theta)\{\delta F(\theta, \tilde{x}(\theta); \varepsilon) + \delta F_x(\theta, \tilde{x}(\theta); \varepsilon)\Delta x(\theta) + \cdots\}\, d\theta.$$

Here, the overdots denote the terms of order ≥ 2 in Δx. Therefore, as it is not difficult to show, the solution

$$\Delta_1 x(t; \varepsilon), \qquad t_1 \leq t \leq t_2,$$
$$\Delta_1 x(t_1; \varepsilon) \equiv 0,$$

of the "truncated" equation (which is linear in Δx)

$$\Delta_1 \dot{x} = F_x(t, \tilde{x}(t))\Delta_1 x + \delta F(t, \tilde{x}(t); \varepsilon) + \delta F_x(t, \tilde{x}(t); \varepsilon)\Delta_1 x \qquad (5.18)$$

differs from $\Delta x(t; \varepsilon)$ by order greater than ε:

$$\|\Delta x(t; \varepsilon) - \Delta_1 x(t; \varepsilon)\| \leq o(\varepsilon), \qquad \frac{o(\varepsilon)}{\varepsilon} \to 0 \qquad (\varepsilon \to 0).$$

The "truncated" equation thus obtained differs from the equation in variations (5.5) by the term $\delta F_x(t, \tilde{x}(t); \varepsilon)\Delta_1 x$ on the right-hand side. If the condition (5.16) is satisfied, then, on the basis of Theorem 5.1, the solution $\delta x(t; \varepsilon)$ of the equation in variations (5.5) with zero initial value differs from $\Delta x(t; \varepsilon)$ by order greater than ε. However, the weaker condition (5.15) does not guarantee us a solution $\delta x(t; \varepsilon)$ with this property. If (5.15) holds we can only guarantee a solution $\Delta_1 x(t; \varepsilon)$ of the truncated equation which differs from $\Delta x(t; \varepsilon)$ by order greater than ε.

We shall now present an example in which the seminorm (5.15) has order ε due to the fact that the functions $\delta F(t, x; \varepsilon)$ and $\delta F_x(t, x; \varepsilon)$ oscillate along the t-axis more and more rapidly as $\varepsilon \to 0$. In this example, the solution $\Delta_1 x(t; \varepsilon)$ of the "truncated" equation (5.18) is "in resonance" with $\delta F_x(t, x; \varepsilon)$

as $\varepsilon \to 0$. For this reason, the term $\delta F_x(t, \tilde{x}(t); \varepsilon)\Delta_1 x(t; \varepsilon)$ turns out to be of order ε, although both of its factors are also of order ε. Therefore, this term can no longer be neglected.*

Here is the example: The initial equation (x is a scalar) is

$$\dot{x} = 0.$$

The initial solution is

$$\tilde{x}(t) \equiv 0, \quad 0 \leq t \leq 1.$$

The family of perturbations is

$$\delta F(t, x; \varepsilon) = x \sin(t/\varepsilon) + \cos(t/\varepsilon).$$

The equation for the perturbation $\Delta x(t; \varepsilon)$ is equation (5.17). The same equation is the "truncated" equation (5.18),

$$\Delta \dot{x} = \cos(t/\varepsilon) + \sin(t/\varepsilon)\Delta x.$$

The equation of variation is

$$\delta \dot{x} = \cos(t/\varepsilon), \quad \delta x(t; \varepsilon) = \varepsilon \sin(t/\varepsilon).$$

The perturbation is

$$\Delta x(t; \varepsilon) = e^{-\varepsilon \cos(t/\varepsilon)} \int_0^t \cos\frac{\theta}{\varepsilon} e^{\varepsilon \cos(\theta/\varepsilon)} d\theta$$

$$= (1 - \varepsilon \cos(t/\varepsilon) + \cdots) \int_0^t \cos(\theta/\varepsilon)(1 + \varepsilon \cos(\theta/\varepsilon) + \cdots) d\theta$$

$$= (1 - \varepsilon \cos(t/\varepsilon) + \cdots)(\varepsilon \sin(t/\varepsilon) + \varepsilon \int_0^t \cos^2(\theta/\varepsilon) d\theta + \cdots)$$

$$= \varepsilon \sin(t/\varepsilon) + \varepsilon \int_0^t \cos^2(\theta/\varepsilon) d\theta + \varepsilon^2 \{\cdots\}.$$

---

*We emphasize that, by definition, the order of the factor $\delta F_x(t, \tilde{x}(t); \varepsilon)$ is equal to the order with respect to ε of the value

$$\|\delta F_x(t, x; \varepsilon)\|_{w, V_{\tilde{x}(t)}}$$

and that the orders of the factor $\Delta_1 x(t; \varepsilon)$ and the product $\delta F_x(t, \tilde{x}(t); \varepsilon)\Delta_1 x(t; \varepsilon)$ are, respectively, equal to the orders of the values

$$\max_{t_1 \leq t \leq t_2} |\Delta_1 x(t; \varepsilon)| \quad \text{and} \quad \max_{t_1 \leq t \leq t_2} |\delta F_x(t, \tilde{x}(t); \varepsilon)\Delta_1 x(t; \varepsilon)|.$$

Therefore,
$$\Delta_2 x(t;\varepsilon) = \Delta x(t;\varepsilon) - \delta x(t;\varepsilon) = \varepsilon \int_0^t \cos^2(\theta/\varepsilon)\, d\theta + \varepsilon^2 \{\cdots\}.$$

The first term in the expansion of $\Delta x(t;\varepsilon)$ in powers of ε, i.e., the term $\varepsilon \sin(t/\varepsilon)$, is in resonance with the function

$$\delta F_x(t, \tilde{x}(t); \varepsilon) = \sin(t/\varepsilon).$$

5.5. On Solutions of Linear Matrix Differential Equations

In this section, we shall construct a solution of a matrix differential equation by the successive approximation method, and we shall study the dependence of the solution on the matrix of the equation coefficients.

Let $\mathscr{I}$ be an arbitrary interval of the t-axis, which can coincide with R. We denote by E_A the linear space of all matrix-valued $n \times n$ functions $A(t)$, $t \in \mathscr{I}$, which are summable on $\mathscr{I}$:

$$\int_{\mathscr{I}} |A(t)|\, dt < \infty.$$

We assume that the space E_A is normed by the norm

$$\|A(t)\|_w = \sup_{t', t'' \in \mathscr{I}} \left| \int_{t'}^{t''} A(t)\, dt \right|.$$

We shall say that a family of matrices $D \subset E_A$ is *integrally uniformly bounded* if the set of values

$$\left\{ \int_{\mathscr{I}} |A(t)|\, dt : A(t) \in D \right\}$$

is bounded. The corresponding family of n-dimensional functions

$$\{F(t,x) = A(t)x : A(t) \in D, x \in R^n\} \subset E_1(\mathscr{I} \times R^n)$$

is, obviously, uniformly Lipschitzian on a neighborhood of an arbitrary compact subset of $\mathscr{I} \times R^n$.

A family $D \subset E_A$ is said to be *integrally equicontinuous* if, for every $\varepsilon > 0$, there exists a positive value $\delta > 0$ such that $|t' - t''| \leq \delta$, $t', t'' \in \mathscr{I}$, implies

$$\left| \int_{t'}^{t''} |A(t)| \, dt \right| \leq \varepsilon \qquad \forall A(t) \in D.$$

For example, if the absolute values of all matrices of a family D are uniformly bounded,

$$|A(t)| \leq C \qquad \forall A(t) \in D,$$

then, obviously, the family D is integrally equicontinuous.

We denote by $C_{\mathscr{I}}^{n^2}$ the complete normed space of matrix-valued, $n \times n$ *continuous functions*

$$X(t), \qquad t \in \mathscr{I},$$

which have finite limits as t tends to the end points of the interval $\mathscr{I}$, and with the norm given by uniform convergence on $\mathscr{I}$:

$$\|X(t)\| = \sup_{t \in \mathscr{I}} |X(t)|.$$

Any matrix-valued absolutely continuous function $X(t)$, $t \in \mathscr{I}$, that satisfies the matrix differential equation

$$\dot{X} = A(t)X, \qquad A(t) \in E_A,$$

almost everywhere on $\mathscr{I}$ is said to be a *solution* of this equation. A solution $\Gamma(t)$, $t \in \mathscr{I}$, of this equation is called a *fundamental matrix* of the equation if $\Gamma(t)$ is a nonsingular matrix for all $t \in \mathscr{I}$. It follows directly from the uniqueness theorem that a necessary and sufficient condition for $\Gamma(t)$ to be nonsingular on $\mathscr{I}$ is that $\Gamma(t)$ be nonsingular at least at a single point. Indeed, if the rank of the matrix $\Gamma(t)$ is less than n at an arbitrary point t_0, then there exists a constant column $x_0 \neq 0$ such that $\Gamma(t_0)x_0 = 0$. It follows that the n-dimensional function $x(t) = \Gamma(t)x_0$, which is a solution of the linear equation

$$\dot{x} = A(t)x,$$

vanishes at $t = t_0$. Therefore, $x(t) \equiv 0$, i.e., the rank of $\Gamma(t)$ is less than n at every point $t \in \mathscr{I}$.

Theorem 5.2. Every matrix equation

$$\dot{X} = A(t)X, \qquad A(t) \in E_A, \tag{5.19}$$

Solutions of Differential Equations: Variation Formula

has a unique solution

$$X(t) = X(t; \tau, X_\tau, A), \qquad t \in \mathscr{I},$$

that satisfies an arbitrary given initial condition

$$X(\tau) = X_\tau, \qquad \tau \in \mathscr{I}.$$

Let there be given an integrally uniformly bounded set $D \subset E_A$ and a solution

$$\hat{X}(t) = X(t; \hat{\tau}, \hat{X}_\tau, \hat{A}), \qquad t \in \mathscr{I},$$

of the equation

$$\dot{X} = \hat{A}(t)X, \qquad \hat{A}(t) \in D.$$

Then, for every $\varepsilon > 0$, there exists a positive value $\delta > 0$ such that, for any initial data (τ, X_τ), $\tau \in \mathscr{I}$, and any matrix $A(t) \in D$ that satisfy the condition

$$|\tau - \hat{\tau}| + |X_\tau - \hat{X}_\tau| + \|A(t) - \hat{A}(t)\|_w \leq \delta,$$

the corresponding solution $X(t; \tau, X_\tau, A)$ satisfies the estimate

$$\|X(t; \tau, X_\tau, A)\theta - X(t; \hat{\tau}, \hat{X}_\tau, \hat{A})\| \leq \varepsilon. \tag{5.20}$$

If we further assume that D is an integrally equicontinuous family, and that $\mathfrak{N}$ is a bounded set of initial values X_τ, then the set of solutions

$$\{X(t; \tau, X_\tau, A): X_\tau \in \mathfrak{N}, A(t) \in D\} \tag{5.21}$$

is a relatively compact (and therefore, bounded) subset of the complete normed space $C_{\mathscr{I}}^{n^2}$.

Proof. We shall use the notation E_σ to represent the normed linear space of points

$$\sigma = (\tau, X_\tau, A(t)) \in R \times E_{X_\tau} \times E_A$$

with the norm

$$\|\sigma\| = |\tau| + |X_\tau| + \|A(t)\|_w.$$

We define a family of mappings $\varphi(X(\cdot), \sigma)$ of the space $C_{\mathscr{I}}^{n^2}$ into itself, which depend on the parameter

$$\sigma = (\tau, X_\tau, A(t)) \in \mathscr{I} \times E_{X_\tau} \times E_A \subset E_\sigma,$$

by the formula

$$X(\cdot) \to \varphi(X(\cdot), \sigma) = X_\tau + \int_\tau^t A(\theta)X(\theta)\, d\theta = Y(t; X(\cdot), \sigma), \qquad t \in J.$$

This mapping indeed transforms $C_{\mathscr{I}}^{n^2}$ into itself. Since every function $X(t) \in C_{\mathscr{I}}^{n^2}$ is bounded by a constant,

$$|Y(t''; X(\cdot), \sigma) - Y(t'; X(\cdot), \sigma)| \leq \text{const} \left| \int_{t'}^{t''} |A(t)| \, dt \right|.$$

It follows that the left-hand side tends to zero as t' and t'' tend to the same end point of the interval $\mathscr{I}$, which proves our assertion.

Every fixed point of the mapping φ is a solution of equation (5.19) with the initial data (τ, X_τ):

$$X_\sigma(t) = X(t; \sigma) = X_\tau + \int_\tau^t A(\theta) X_\sigma(\theta) \, d\theta, \qquad t \in \mathscr{I}.$$

Therefore, we shall obtain the existence of a unique solution and the relation (5.20) (the continuous dependence) as a corollary of the fixed-point theorem (Theorem 4.1) if we prove the following: First, the iterations $\varphi^{(p)}(X(\cdot), \sigma)$ depend continuously on σ for a fixed $X(\cdot)$ and for σ which ranges over a set of the form

$$\mathscr{I} \times E_{X_\tau} \times D \subset E_\sigma, \qquad (5.22)$$

where $D \subset E_A$ is an integrally equicontinuous family of matrices $A(t)$. Second, these iterations with p sufficiently large form a contraction family of mappings on the set (5.22).

Since the proof of these properties of the mapping φ would be a repetition of the proof of Theorem 4.2 (with some simplifications), we shall assume that the properties are proved, and we pass directly to the proof of the relative compactness of the set (5.21), i.e., of the fact that the closure of the set (5.21) in $C_{\mathscr{I}}^{n^2}$ is compact. This is the last assertion of the theorem.

First, we shall prove the uniform boundedness of the family (5.21). We write the following series of inequalities:

$$|X(t)| = |X(t; \tau, X_\tau, A)| \leq |X_\tau| + \left| \int_\tau^t |A(\theta_1)| \cdot |X(\theta_1)| \, d\theta_1 \right|$$

$$\leq |X_\tau| \left(1 + \left| \int_\tau^t |A(\theta_1)| \, d\theta_1 \right| \right) + \left| \int_\tau^t |A(\theta_1)| \, d\theta_1 \int_\tau^{\theta_1} |A(\theta_2)| \cdot |X(\theta_2)| \, d\theta_2 \right| \leq \cdots$$

Solutions of Differential Equations: Variation Formula

$$\leqslant |X_\tau| \left\{ 1 + \left| \int_\tau^t |A(\theta_1)|\, d\theta_1 \right| + \left| \int_\tau^t |A(\theta_1)|\, d\theta_1 \int_\tau^{\theta_1} |A(\theta_2)|\, d\theta_2 \right| + \cdots \right.$$

$$+ \left| \int_\tau^t |A(\theta_1)|\, d\theta_1 \int_\tau^{\theta_1} |A(\theta_2)|\, d\theta_2 \cdots \int_\tau^{\theta_{p-1}} |A(\theta_p)|\, d\theta_p \right| \bigg\}$$

$$+ \left| \int_\tau^t |A(\theta_1)|\, d\theta_1 \cdots \int_\tau^{\theta_{p-1}} |A(\theta_p)|\, d\theta_p \int_\tau^{\theta_p} |A(\theta_{p+1})| \cdot |X(\theta_{p+1})|\, d\theta_{p+1} \right|.$$

We make use of the following formula, proved in Chapter 4:

$$\int_\tau^t |A(\theta_1)|\, d\theta_1 \int_\tau^{\theta_1} |A(\theta_2)|\, d\theta_2 \cdots \int_\tau^{\theta_{p-1}} |A(\theta_p)|\, d\theta_p = \frac{1}{p!} \left(\int_\tau^t |A(\theta)|\, d\theta \right)^p$$

Taking into account that

$$\int_{\mathscr{I}} |A(\theta)|\, d\theta \leqslant C \qquad \forall A(t) \in D,$$

we can write

$$|X(t)| \leqslant |X_\tau| \left(1 + C + \frac{C^2}{2!} + \cdots + \frac{C^p}{p!} \right) + \|X(t)\| \frac{C^{p+1}}{(p+1)!}.$$

Passing to the limit as $p \to \infty$, we obtain the estimate

$$\|X(t)\| \leqslant |X_\tau| e^C.$$

This estimate proves the uniform boundedness of the norms $\|X(t; \tau, X_\tau, A)\|$, since the set $\mathfrak{N}$ of initial values X_τ is bounded by assumption.

The equicontinuity of the family (5.21) follows easily from the integral uniform continuity of the family D. Indeed, independently of the choice of $X_\tau \in \mathfrak{N}$ and $A(t) \in D$, we have

$$|X(t'; \tau, X_\tau, A) - X(t''; \tau, X_\tau, A)| \leqslant \left| \int_{t'}^{t''} |A(t)| \cdot |X(t; \tau, X_\tau, A)|\, dt \right|$$

$$\leqslant \|X(t; \tau, X_\tau, A)\| \cdot \left| \int_{t'}^{t''} |A(t)|\, dt \right| \to 0$$

as $|t'-t''|\to 0$. This is so for the following reasons: As we have just proved, the norms $\|X(t;\tau, X_\tau, A)\|$ are uniformly bounded. Also, it follows from the integral uniform continuity of the family D that

$$\int_{t'}^{t''} |A(t)|\, dt \to 0.$$

Now the relative compactness of the family (5.21) follows from the Arzelà theorem, which is well known from analysis.

6

The Varying of Trajectories in Convex Control Problems

The results of the preceding two chapters allow us to define a very natural method for varying the trajectories in convex control problems. This method has a number of properties which make it an important investigative tool in control problems. This chapter is devoted to the description of the method and the proof of its basic properties.

6.1. Variations of Generalized Controls and the Corresponding Variations of the Controlled Equation

We return to the convex control problem

$$\dot{x} = \langle \mu_t, f(t, x, u) \rangle, \quad \mu_t \in \mathfrak{M}_U.$$

Let $\bar{\mu}_t$ be an arbitrary generalized control, and let $\bar{x}(t)$, $t_1 \leq t \leq t_2$, be a trajectory of the equation

$$\dot{x} = \langle \bar{\mu}_t, f(t, x, u) \rangle = F(t, x). \tag{6.1}$$

The function $F(t, x)$ is defined on the entire space $R \times R^n$, continuously differentiable with respect to x, measurable in t, and bounded on any compact set $K \subset R \times R^n$. The boundness follows from the following reasons: The probability measure $\bar{\mu}_t$ is concentrated on a bounded set

$$N \subset U \subset R^r,$$

which does not depend on $t \in R$, and the function $f(t, x, u)$ is continuous. Hence

$$\sup_{(t,x) \in K} |F(t, x)| \leq \sup_{(t,x) \in K} \int_N |f(t, x, u)| \, d\tilde{\mu}_t$$

$$\leq \sup_{\substack{(t,x) \in K \\ u \in N}} |f(t, x, u)| \int_N d\tilde{\mu}_t = \sup_{\substack{(t,x) \in K \\ u \in N}} |f(t, x, u)|. \quad (6.2)$$

Therefore, we have $\alpha(t, x) F(t, x) \in E_1$ for any scalar-valued continuously differentiable function $\alpha(t, x)$ with compact support, i.e.,

$$F(t, x) \in E_1(R \times R^n).$$

Any difference

$$\delta \mu_t = \mu_t - \tilde{\mu}_t, \quad \mu_t \in \mathfrak{M}_U,$$

will be called a *variation*, or a *perturbation* of the generalized control $\tilde{\mu}_t$. The set of all perturbations of the control $\tilde{\mu}_t$ will be denoted by $\delta \mathfrak{M}_{\tilde{\mu}_t}$. The convexity of $\delta \mathfrak{M}_{\tilde{\mu}_t}$ follows immediately from the convexity of $\mathfrak{M}_U$:

$$\lambda_1(\mu_t^{(1)} - \tilde{\mu}_t) + \lambda_2(\mu_t^{(2)} - \tilde{\mu}_t) = \lambda_1 \mu_t^{(1)} + \lambda_2 \mu_t^{(2)} - \tilde{\mu}_t \in \delta \mathfrak{M}_{\tilde{\mu}_t}$$
$$\forall \lambda_1, \lambda_2 \geq 0, \quad \lambda_1 + \lambda_2 = 1.$$

To every perturbation $\delta \mu_t \in \delta \mathfrak{M}_{\tilde{\mu}_t}$, there corresponds, in a natural way, a perturbation on the right-hand side of equation (6.1). Substituting a *perturbed (varied) control* $\tilde{\mu}_t + \delta \mu_t$ for $\tilde{\mu}_t$ in (6.1), we obtain the *perturbed (varied) equation*

$$\dot{x} = \langle \tilde{\mu}_t + \delta \mu_t, f(t, x, u) \rangle = \langle \tilde{\mu}_t, f(t, x, u) \rangle + \langle \delta \mu_t, f(t, x, u) \rangle.$$

We can see that the *perturbation on the right-hand side*

$$\langle \delta \mu_t, f(t, x, u) \rangle = \delta F(t, x)$$

depends linearly (more precisely, the dependence is "affine") on a control perturbation $\delta \mu_t$. One can combine two perturbations $\delta \mu_t^{(1)}$ and $\delta \mu_t^{(2)}$ with coefficients λ_1 and λ_2 that satisfy the conditions $\lambda_1 \geq 0, \lambda_2 \geq 0$, and $\lambda_1 + \lambda_2 = 1$. This makes the convex control problem much more flexible for investigations than the initial control problem with ordinary controls. In particular, this fact together with the approximation lemma play the decisive role in the proof of the maximum principle in the next chapter.

Varying of Trajectories in Convex Control Problems

Every perturbation $\langle \delta\mu_t, f(t, x, u) \rangle$ belongs to $E_1(R \times R^n)$ by the same argument used to show that $F(t, x)$ does.

The equation of variation for (6.1) along the trajectory $\tilde{x}(t)$, $t_1 \leq t \leq t_2$, has the form

$$\delta\dot{x} = F_x(t, \tilde{x}(t))\delta x + \langle \delta\mu_t, f(t, \tilde{x}(t), u) \rangle.$$

Retaining the notation of Chapter 5, we can write a variation of the solution $\tilde{x}(t)$ with zero initial value in the form

$$\delta x(t; \delta\mu_\theta) = \Gamma(t) \int_{t_1}^{t} G(\theta) \langle \delta\mu_\theta, f(\theta, \tilde{x}(\theta), u) \rangle \, d\theta.$$

A family of perturbations

$$\{\delta\mu_t(\sigma) : \sigma \in \Sigma\},$$

where Σ is an arbitrary set of indices, will be called *finite* if all measures $\delta\mu_t(\sigma)$ are concentrated on a single bounded subset of the space R^r which depends neither on $t \in R$ nor on $\sigma \in \Sigma$.

Assertion 6.1. For every finite family of perturbations, the corresponding family of functions

$$\{\langle \delta\mu_t(\sigma), f(t, x, u) \rangle = \delta F(t, x; \sigma) : \sigma \in \Sigma\}$$

is uniformly Lipschitzian on a neighborhood of any compact set $K \subset R \times R^n$.

Proof. We take any function $\alpha_{V_K}(t, x)$, where V_K is an arbitrary compact neighborhood of the set K. Let I be the projection of the support of the function $\alpha_{V_K}(t, x)$ onto the t-axis, and let X be its projection onto the subspace R^n.

As in the proof of inequality (6.2), we prove that all the functions

$$|\delta F(t, x; \sigma)| \quad \text{and} \quad |\delta F_x(t, x; \sigma)| \quad \text{with } \sigma \in \Sigma$$

are bounded by a single constant C on the support of the function $\alpha_{V_K}(t, x)$. Therefore, the following inequalities hold:

$$\|\delta F(t, x; \sigma)\|_{1, V_K} \leq \int_R \max_{x \in R^n} \left\{ |\alpha_{V_K}(t, x) \delta F(t, x; \sigma)| + \left| \frac{\partial}{\partial x} \alpha_{V_K}(t, x) \delta F(t, x; \sigma) \right| \right\} dt$$

$$\leq \int_I \max_{x \in X} \left\{ |\delta F(t, x; \sigma)| + |\delta F_x(t, x; \sigma)| + \left| \frac{\partial}{\partial x} \alpha_{V_K}(t, x) \right| \cdot |\delta F(t, x; \sigma)| \right\} dt$$

$$\leq C \int_I \left(2 + \max_{x \in R^n} \left| \frac{\partial}{\partial x} \alpha_{V_K}(t, x) \right| \right) dt.$$

Assertion 6.1 now follows from these inequalities and from Assertion 5.1.

The following assertion establishes the connection between the weak convergence of generalized controls and the seminorm $\|\cdot\|_w$:

Assertion 6.2. Let a sequence of perturbations which depend on a parameter $\sigma \in \Sigma$,

$$\delta \mu_t^{(i)}(\sigma), \qquad \sigma \in \Sigma, \quad i = 1, 2, \ldots,$$

converge weakly to zero as $i \to \infty$, uniformly with respect to $\sigma \in \Sigma$. This means that for an arbitrary continuous function $g(t, u)$ on $R \times R^n$ with compact support (see Chapter 2), we have

$$\int_R \langle \delta \mu_t^{(i)}(\sigma), g(t, u) \rangle \, dt \to 0 \qquad (i \to \infty)$$

uniformly with respect to $\sigma \in \Sigma$. Let all measures $\delta \mu_t^{(i)}(\sigma)$, $t \in R$, $i = 1, 2, \ldots$, $\sigma \in \Sigma$, be concentrated on a single bounded set

$$N \subset U \subset R^r.$$

Then for an arbitrary compact set $K \subset R \times R^n$, we have

$$\|\langle \delta \mu_t^{(i)}(\sigma), f(t, x, u) \rangle\|_{w,K} \to 0 \qquad (i \to \infty),$$

where the convergence to zero is also uniform with respect to $\sigma \in \Sigma$.

Proof. We choose a function $\alpha_K(t, x)$. Let $\hat{K}$ be its support. We use the notation

$$g_\sigma^{(i)}(t', t'', x) = \left| \int_{t'}^{t''} \langle \delta \mu_t^{(i)}(\sigma), \alpha_K(t, x) f(t, x, u) \rangle \, dt \right|.$$

Since

$$\|\langle \delta \mu_t^{(i)}(\sigma), f(t, x, u) \rangle\|_{w,K} \leq \max_{\substack{t', t'' \in R \\ x \in R^n}} g_\sigma^{(i)}(t', t'', x) \qquad \forall \sigma \in \Sigma,$$

we shall obtain the assertion if we show that

$$g_\sigma^{(i)}(t', t'', x) \to 0 \qquad (i \to \infty)$$

uniformly with respect to $t', t'' \in R$, $x \in R^n$, and $\sigma \in \Sigma$.

Assuming the contrary, we can find the following: a positive number $\eta > 0$, an increasing sequence of integers $i_1, i_2, \ldots, i_k, \ldots$, convergent sequences (in view of the compactness of the support $\hat{K}$)

$$t'^{(k)} \to \hat{t}', \quad t''^{(k)} \to \hat{t}'', \quad x^{(k)} \to \hat{x} \qquad (k \to \infty),$$

and a sequence of points $\sigma_k \in \Sigma$ such that

$$g_{\sigma_k}^{(i_k)}(t'^{(k)}, t''^{(k)}, x^{(k)}) \geq \eta > 0 \qquad \forall k = 1, 2, \ldots.$$

We consider the inequality

$$\left| g_{\sigma_k}^{(i_k)}(\hat{t}', \hat{t}'', \hat{x}) - g_{\sigma_k}^{(i_k)}(t'^{(k)}, t''^{(k)}, x^{(k)}) \right|$$

$$\leq \left| \int_{\hat{t}'}^{\hat{t}''} |\langle \delta\mu_t^{(i_k)}(\sigma_k), \alpha_K(t, \hat{x}) f(t, \hat{x}, u) - \alpha_K(t, x^{(k)}) f(t, x^{(k)}, u) \rangle| \, dt \right|$$

$$+ \left| \int_{\hat{t}''}^{t''^{(k)}} |\langle \delta\mu_t^{(i_k)}(\sigma_k), \alpha_K(t, x^{(k)}) f(t, x^{(k)}, u) \rangle| \, dt \right|$$

$$+ \left| \int_{t'^{(k)}}^{\hat{t}'} |\langle \delta\mu_t^{(i_k)}(\sigma_k), \alpha_K(t, x^{(k)}) f(t, x^{(k)}, u) \rangle| \, dt \right|. \qquad (6.3)$$

We shall show that each of the terms on the right-hand side and the function $g_{\sigma_k}^{(i_k)}(\hat{t}', \hat{t}'', \hat{x})$ tend to zero as $k \to \infty$. Therefore, for large k, we shall have

$$0 < \eta/2 < \left| g_{\sigma_k}^{(i_k)}(\hat{t}', \hat{t}'', \hat{x}) - g_{\sigma_k}^{(i_k)}(t'^{(k)}, t''^{(k)}, x^{(k)}) \right| < \eta/2,$$

i.e., we shall obtain the contradiction which proves our assertion.

All measures $\delta\mu_t^{(i)}(\sigma)$ are concentrated on the bounded set $N \subset R^r$, and their norms are equal to or less than 2, as the differences between two probability measures. Therefore, by Assertion 2.2 we have that for any *fixed* t', t'', and x,

$$g_\sigma^{(i)}(t', t'', x) = \int_{t'}^{t''} \langle \delta\mu_t^{(i)}(\sigma), f(t, x, u) \rangle \, dt \to 0 \qquad (i \to \infty)$$

uniformly with respect to $\sigma \in \Sigma$. In particular,

$$g^{(ik)}_{\sigma_k}(\hat{t}', \hat{t}'', \hat{x}) \to 0 \quad (k \to \infty).$$

We now turn to the terms on the right-hand side of the inequality (6.3). In the same way that we proved the formula (6.2), we prove that, for all $(t, x) \in \hat{K}$ [and, therefore, for all $(t, x) \in R \times R^n$],

$$|\langle \delta\mu_t^{(i)}(\sigma), \alpha_K(t, x) f(t, x, u) \rangle| \leq \text{const}.$$

Therefore,

$$\left| \int_{t'^{(k)}}^{\hat{t}'} |\langle \delta\mu_t^{(ik)}(\sigma_k), \alpha_K(t, x^{(k)}) f(t, x^{(k)}, u) \rangle| \, dt \right| \leq \text{const} \left| \int_{t'^{(k)}}^{\hat{t}'} dt \right| \to 0,$$

$$\left| \int_{\hat{t}''}^{t''^{(k)}} |\langle \delta\mu_t^{(ik)}(\sigma_k), \alpha_K(t, x^{(k)}) f(t, x^{(k)}, u) \rangle| \, dt \right| \leq \text{const} \left| \int_{\hat{t}''}^{t''^{(k)}} dt \right| \to 0.$$

Since the set N is bounded, since $x^{(k)} \to \hat{x}$, and since the function $\alpha_K(t, x) f(t, x, u)$ is continuous,

$$\sup_{u \in N} |\alpha_K(t, \hat{x}) f(t, \hat{x}, u) - \alpha_K(t, x^{(k)}) f(t, x^{(k)}, u)| \to 0$$

as $k \to \infty$ uniformly with respect to $t \in [\hat{t}', \hat{t}'']$. Hence we finally have

$$\left| \int_{\hat{t}'}^{\hat{t}''} |\langle \delta\mu_t^{(ik)}(\sigma_k), \alpha_K(t, \hat{x}) f(t, \hat{x}, u) - \alpha_K(t, x^{(k)}) f(t, x^{(k)}, u) \rangle| \, dt \right|$$

$$\leq \left| \int_{\hat{t}'}^{\hat{t}''} dt \int_N |\alpha_K(t, \hat{x}) f(t, \hat{x}, u) - \alpha_K(t, x^{(k)}) f(t, x^{(k)}, u)| \, d\mu_t^{(ik)} \right|$$

$$\leq 2 \left| \int_{\hat{t}'}^{\hat{t}''} \sup_{u \in N} |\alpha_K(t, \hat{x}) f(t, \hat{x}, u) - \alpha_K(t, x^{(k)}) f(t, x^{(k)} u)| \, dt \right| \to 0.$$

If we replace the weak convergence in Assertion 6.2 by the strong convergence, then we obtain the following assertion:

Assertion 6.3. Let a sequence of perturbations

$$\delta\mu_t^{(i)}(\sigma), \sigma \in \Sigma, \quad i = 1, 2, \ldots,$$

converge strongly to zero uniformly with respect to $\sigma \in \Sigma$, i.e.,

$$\int_R \|\delta\mu_t^{(i)}(\sigma)\| \, dt \to 0 \quad (i \to \infty)$$

Varying of Trajectories in Convex Control Problems   105

uniformly with respect to $\sigma \in \Sigma$. Also, assume that all measures $\delta\mu_t^{(i)}(\sigma)$ are concentrated on a single bounded set $N \subset R^r$. Then for an arbitrary compact set $K \subset R \times R^n$, we have

$$\|\langle \delta\mu_t^{(i)}(\sigma), f(t, x, u)\rangle\|_{1,K} \to 0 \qquad (i \to \infty)$$

uniformly with respect to $\sigma \in \Sigma$.

Proof. We denote

$$\langle \delta\mu_t^{(i)}(\sigma), f(t, x, u)\rangle = \delta F^{(i)}(t, x; \sigma).$$

We choose a function $\alpha_K(t, x)$. Let $\hat{K}$ be its support. The following obvious estimates hold uniformly with respect to $\sigma \in \Sigma$:

$$\sup_{x \in R^n} \left\{ |\alpha_K(t, x)\delta F^{(i)}(t, x; \sigma)| + \left|\frac{\partial}{\partial x} \alpha_K(t, x)\delta F^{(i)}(t, x; \sigma)\right| \right\}$$

$$\leq \sup_{\substack{(\theta,x) \in R \times R^n \\ u \in N}} \left\{ |\alpha_K(\theta, x)f(\theta, x, u)| + \left|\frac{\partial}{\partial x} \alpha_K(\theta, x)f(\theta, x, u)\right| \right\} \|\delta\mu_t^{(i)}(\sigma)\|.$$

Thus, we obtain the relations

$$\sup_{\sigma \in \Sigma} \|\delta F^{(i)}(t, x; \sigma)\|_{1,K}$$

$$\leq \sup_{\sigma \in \Sigma} \int_R \sup_{x \in R^n} \left\{ |\alpha_K(t, x)\delta F^{(i)}(t, x; \sigma)| + \left|\frac{\partial}{\partial x} \alpha_K(t, x)\delta F^{(i)}(t, x; \sigma)\right| \right\} dt$$

$$\leq \sup_{\substack{(t,x) \in R \times R^n \\ u \in N}} \left\{ |\alpha_K(t, x)f(t, x, u)| + \left|\frac{\partial}{\partial x} \alpha_K(t, x)f(t, x, u)\right| \right\} \sup_{\sigma \in \Sigma} \int_R \|\delta\mu_t^{(i)}(\sigma)\| \, dt \to 0$$

$$(i \to \infty),$$

which prove the assertion.

To conclude this section, we shall describe an important method for the construction of a family of variations of a generalized control. In the next section, we will construct a family of varied trajectories with the aid of this method. This family is used in Chapter 7 in the proof of the maximum principle.

Let there be given p perturbations of the control $\tilde{\mu}_t$

$$\delta\mu_t^{(1)}, \ldots, \delta\mu_t^{(p)} \in \delta\mathfrak{M}_{\tilde{\mu}_t}$$

which *vanish outside of an (arbitrary) interval* of the t-axis $I \subset R$:

$$\delta\mu_t^{(k)} = 0 \qquad \forall t \notin I \quad \text{and} \quad \forall k = 1, 2, \ldots, p.$$

For any fixed nonnegative numbers $\lambda_1 \geq 0, \ldots, \lambda_p \geq 0$, the family of measures

$$\delta\mu_t(\varepsilon, \lambda) = \varepsilon \sum_{k=1}^{p} \lambda_k \delta\mu_t^{(k)}, \qquad t \in R, \qquad \lambda = (\lambda_1, \ldots, \lambda_p), \qquad \varepsilon \geq 0,$$

is a family of perturbations of the control $\tilde{\mu}_t$ for all sufficiently small $\varepsilon \geq 0$. Indeed, the set $\delta\mathfrak{M}_{\tilde{\mu}_t}$ is convex and contains the zero perturbation, $\delta\mu_t = 0 \in \delta\mathfrak{M}_{\tilde{\mu}_t}$, so that for $1 - \varepsilon \sum_{k=1}^{p} \lambda_k \geq 0$, we have

$$\delta\mu_t(\varepsilon, \lambda) = \sum_{k=1}^{p} \varepsilon \lambda_k \delta\mu_t^{(k)} + \left(1 - \sum_{k=1}^{p} \varepsilon \lambda_k\right) 0 \in \delta\mathfrak{M}_{\tilde{\mu}_t}.$$

The upper limit for ε depends, of course, on λ. Assume that λ ranges over a bounded set $\Lambda \subset R^p$, where

$$\lambda = (\lambda_1, \ldots, \lambda_p) \geq 0 \qquad \forall \lambda \in \Lambda,$$

i.e., $\lambda_1 \geq 0, \ldots, \lambda_p \geq 0$. Then, for all $\lambda \in \Lambda$, a single positive upper limit $\varepsilon_0 > 0$ of the range of ε can be given, and we obtain the family of perturbations

$$\left\{\delta\mu_t(\varepsilon, \lambda) = \varepsilon \sum_{k=1}^{p} \lambda_k \delta\mu_t^{(k)}: \lambda \in \Lambda, 0 \leq \varepsilon \leq \varepsilon_0\right\} \subset \delta\mathfrak{M}_{\tilde{\mu}_t}. \tag{6.4}$$

Obviously, this family is finite. Since the norms $\|\delta\mu_t^{(k)}\| \leq 2$, we have

$$\int_R \|\delta\mu_t(\varepsilon, \lambda) - \delta\mu_t(\hat{\varepsilon}, \hat{\lambda})\| \, dt = \int_R \left\|\sum_{k=1}^{p} (\varepsilon\lambda_k - \hat{\varepsilon}\hat{\lambda}_k)\delta\mu_t^{(k)}\right\| dt$$

$$\leq \sum_{k=1}^{p} |\varepsilon\lambda_k - \hat{\varepsilon}\hat{\lambda}_k| \int_I \|\delta\mu_t^{(k)}\| \, dt \to 0,$$

as $(\varepsilon, \lambda) \to (\hat{\varepsilon}, \hat{\lambda})$, i.e., the family of perturbations (6.4) depends strongly (and therefore weakly) continuously on $(\varepsilon, \lambda) \in [0, \varepsilon_0] \times \Lambda$.

Further, the family $\delta\mu_t(\varepsilon, \lambda)$ converges to zero strongly (and therefore weakly) as $\varepsilon \to 0$, uniformly with respect to $\lambda \in \Lambda$. This follows from the estimate,

$$\|\delta\mu_t(\varepsilon, \lambda)\| \leq \varepsilon \cdot \text{const} \qquad \forall t \in R \quad \text{and} \quad \forall \lambda \in \Lambda,$$

which holds by virtue of the boundedness of Λ and because $\delta\mu_t(\varepsilon, \lambda) = 0$ for t with large absolute values (for $t \notin I$).

6.2. Variations of Trajectories

The following assertion is a direct corollary to Assertion 6.1, to Theorem 4.4 on the continuous dependence of solutions, and to Theorem 5.1 on variations of solutions:

Assertion 6.4. Let

$$\{\delta\mu_t(\sigma):\ \sigma\in\Sigma\}\subset\delta\mathfrak{M}_{\tilde\mu_t}$$

be a finite family of perturbations of the control $\tilde\mu_t$, and let $V_{\tilde x(t)}$ be an arbitrary compact neighborhood of a solution $\tilde x(t)$, $t_1 \leqslant t \leqslant t_2$, of the equation

$$\dot x = \langle \tilde\mu_t, f(t,x,u)\rangle = F(t,x).$$

Then there exists a positive number $\varepsilon_1 > 0$ such that, for all $\varepsilon \in [0,\varepsilon_1]$ and for any $\delta\mu_t(\sigma)$ that satisfies the condition

$$\|\langle\delta\mu_t(\sigma), f(t,x,u)\rangle\|_{w,V_{\tilde x(t)}} \leqslant \varepsilon, \tag{6.5}$$

the perturbed equation

$$\dot x = F(t,x) + \langle\delta\mu_t(\sigma), f(t,x,u)\rangle$$

has the solution

$$x(t) = x(t;\delta\mu_\theta(\sigma)),\quad t_1\leqslant t\leqslant t_2,\quad \sigma\in\Sigma,\quad x(t_1;\delta\mu_\theta(\sigma))\equiv\tilde x(t_1),$$

that satisfies the estimate

$$\max_{t\in[t_1,t_2]}|x(t;\delta\mu_\theta(\sigma))-\tilde x(t)| = \|x(t;\delta\mu_\theta(\sigma))-\tilde x(t)\| \leqslant O(\varepsilon),$$

where

$$O(\varepsilon)\to 0\quad(\varepsilon\to 0),\quad O(0)=0.$$

If the condition (6.5) is replaced by a stronger condition

$$\|\langle\delta\mu_t(\sigma), f(t,x,u)\rangle\|_{1,V_{\tilde x(t)}} \leqslant \varepsilon,$$

then the following estimate holds:

$$\left\|x(t;\delta\mu_\theta(\sigma))-\tilde x(t)-\Gamma(t)\int_{t_1}^{t} G(\theta)\langle\delta\mu_\theta(\sigma), f(\theta,\tilde x(\theta), u)\rangle\,d\theta\right\| \leqslant o(\varepsilon),$$

where

$$\frac{o(\varepsilon)}{\varepsilon}\to 0\quad(\varepsilon\to 0).$$

Assertion 6.5. Assume that an arbitrary finite family of perturbations

$$\{\delta\mu_t(\sigma): \sigma \in \Sigma\},$$

where Σ is a metric space, depends weakly continuously on a parameter $\sigma \in \Sigma$, i.e., $\sigma_i \to \hat{\sigma}(i \to \infty)$ implies that the difference $\delta\mu_t(\sigma_i) - \delta\mu_t(\hat{\sigma})$ converges weakly to zero. Furthermore, assume that, for every $\sigma \in \Sigma$, the equation

$$\dot{x} = F(t, x) + \langle \delta\mu_t(\sigma), f(t, x, u) \rangle$$

has the solution

$$x(t; \sigma), \quad t_1 \leq t \leq t_2, \quad \sigma \in \Sigma,$$
$$x(t_1; \sigma) \equiv x_1.$$

Then the function of two variables $x(t; \sigma)$ depends continuously on $(t, \sigma) \in [t_1, t_2] \times \Sigma$.

Proof. Let $\sigma_i \to \hat{\sigma}$ $(i \to \infty)$. We denote by $V_{x(t;\hat{\sigma})}$ an arbitrary compact neighborhood of the curve $x(t; \hat{\sigma})$, $t_1 \leq t \leq t_2$.

According to Assertion 6.2,

$$\|\langle \delta\mu_t(\sigma_i) - \delta\mu_t(\hat{\sigma}), f(t, x, u) \rangle\|_{w, V_{x(t;\hat{\sigma})}} \to 0 \quad (i \to \infty).$$

Also, it follows from Assertion 6.1 that the set of functions

$$\{\langle \delta\mu_t(\sigma_i) - \delta\mu_t(\hat{\sigma}), f(t, x, u) \rangle : i = 1, 2, \ldots\}$$

is uniformly Lipschitzian on a neighborhood of any compact subset of $R \times R^n$. Therefore, we conclude on the basis of Theorem 4.4 on the continuous dependence of solutions that the sequence of solutions $x(t; \sigma_i)$, $t_1 \leq t \leq t_2$, of the equations

$$\dot{x} = F(t, x) + \langle \delta\mu_t(\sigma_i), f(t, x, u) \rangle$$
$$= F(t, x) + \langle \delta\mu_t(\hat{\sigma}), f(t, x, u) \rangle + \langle \delta\mu_t(\sigma_i) - \delta\mu_t(\hat{\sigma}), f(t, x, u) \rangle, \quad i = 1, 2, \ldots,$$

satisfies the condition

$$\max_{t \in [t_1, t_2]} |x(t; \sigma_i) - x(t; \hat{\sigma})| \to 0 \quad (i \to \infty).$$

Hence our assertion follows.

We shall now prove two basic theorems on the construction of families of varied trajectories in the convex control problem. We shall make use of these theorems in the proof of the maximum principle.

It is assumed in the statements of these theorems that the equation

$$\dot{x} = \langle \tilde{\mu}_t, f(t, x, u) \rangle = F(t, x)$$

and a solution of this equation

$$\tilde{x}(t), \quad t_1 \leqslant t \leqslant t_2,$$

are fixed. Moreover, we assume in Theorem 6.2 that the set Σ of values of the parameter is a metric space.

Theorem 6.1. For sufficiently small $\varepsilon \geqslant 0$ (e.g., for $0 \leqslant \varepsilon \leqslant \varepsilon_1$, $\varepsilon_1 > 0$), the family of perturbations (6.4) generates the family of varied solutions

$$x(t; \delta\mu_\theta(\varepsilon, \lambda)) = x(t; \varepsilon, \lambda), \quad t_1 \leqslant t \leqslant t_2, \quad \lambda \in \Lambda,$$
$$x(t_1; \varepsilon, \lambda) \equiv \tilde{x}(t_1)$$

of the perturbed equation

$$\dot{x} = F(t, x) + \langle \delta\mu_t(\varepsilon, \lambda), f(t, x, u) \rangle = F(t, x) + \varepsilon \sum_{k=1}^{p} \lambda_k \langle \delta\mu_t^{(k)}, f(t, x, u) \rangle, \quad \lambda \in \Lambda.$$

The function $x(t; \varepsilon, \lambda)$ depends continuously on

$$(t, \varepsilon, \lambda) \in [t_1, t_2] \times [0, \varepsilon_1] \times \Lambda$$

and can be represented in the form

$$x(t; \varepsilon, \lambda) = \tilde{x}(t) + \varepsilon \sum_{k=1}^{p} \lambda_k \delta x(t; \delta\mu_\theta^{(k)}) + \Delta_2 x(t; \varepsilon, \lambda),$$

where $\delta x(t; \delta\mu_\theta^{(k)})$ is the solution of the equation of variation

$$\delta \dot{x} = F_x(t, \tilde{x}(t))\delta x + \langle \delta\mu_t^{(k)}, f(t, \tilde{x}(t), u) \rangle$$

with zero initial condition

$$\delta x(t_1; \delta\mu_\theta^{(k)}) = 0,$$

and $\Delta_2 x(t; \varepsilon, \lambda)$ satisfies the condition

$$|\Delta_2 x(t; \varepsilon, \lambda)| \leqslant o(\varepsilon) \quad \forall (t, \lambda) \in [t_1, t_2] \times \Lambda, \quad \frac{o(\varepsilon)}{\varepsilon} \to 0 \quad (\varepsilon \to 0).$$

Proof. It follows from Assertion 6.3 that for any compact set $K \subset R \times R^n$, we have

$$\|\langle \delta\mu_t(\varepsilon, \lambda), f(t, x, u) \rangle\|_{1, K} \to 0 \quad (\varepsilon \to 0),$$

since the finite family of perturbations $\delta\mu_t(\varepsilon, \lambda)$ converges strongly to zero as $\varepsilon \to 0$ (uniformly with respect to $\lambda \in \Lambda$). Therefore, the theorem is a direct corollary to Assertions 6.4 and 6.5.

Theorem 6.2. Assume that a finite and weakly continuous family of perturbations

$$\{\delta\mu_t(\sigma): \sigma \in \Sigma\}$$

satisfies the condition

$$\|\langle\delta\mu_t(\sigma), f(t, x, u)\rangle\|_{w, V_{\tilde{x}(t)}} \leq \varepsilon_0 \qquad \forall \sigma \in \Sigma, \quad \varepsilon_0 > 0,$$

where $V_{\tilde{x}(t)}$ is an arbitrary compact neighborhood of the curve $\tilde{x}(t)$, $t_1 \leq t \leq t_2$. Furthermore, let the sequence of perturbations $\delta\mu_t^{(i)}(\sigma)$, $i = 1, 2, \ldots$, which depend weakly continuously on $\sigma \in \Sigma$, converge weakly to $\delta\mu_t(\sigma)$ as $i \to \infty$, uniformly with respect to $\sigma \in \Sigma$. Here, we assume that all measures $\delta\mu_t^{(i)}(\sigma)$, $t \in R$, $\sigma \in \Sigma$, $i = 1, 2, \ldots$, are concentrated on a single bounded subset of R^r. Then, for a sufficiently small $\varepsilon_0 > 0$, the equation

$$\dot{x} = F(t, x) + \langle\delta\mu_t^{(i)}(\sigma), f(t, x, u)\rangle$$

generates, for all i sufficiently large, a family of solutions

$$x^{(i)}(t; \sigma), \quad t_1 \leq t \leq t_2, \quad \sigma \in \Sigma,$$
$$x^{(i)}(t_1; \sigma) \equiv \tilde{x}(t_1), \tag{6.6}$$

with the following property: As $i \to \infty$, this family tends to the family of solutions

$$x(t; \sigma), \quad t_1 \leq t \leq t_2, \quad \sigma \in \Sigma,$$
$$x(t_1; \sigma) \equiv \tilde{x}(t_1), \tag{6.7}$$

of the equation

$$\dot{x} = F(t, x) + \langle\delta\mu_t(\sigma), f(t, x, u)\rangle$$

in the norm of uniform convergence on $[t_1, t_2]$. Moreover, the convergence is uniform with respect to $\sigma \in \Sigma$. The functions $x(t; \sigma)$ and $x^{(i)}(t; \sigma)$ depend continuously on $(t, \sigma) \in [t_1, t_2] \times \Sigma$.

Proof. It follows from the weak convergence $\delta\mu_t^{(i)}(\sigma) \to \delta\mu_t(\sigma)$ and from Assertion 6.2 that the following inequality holds for large i:

$$\|\langle\delta\mu_t^{(i)}(\sigma), f(t, x, u)\rangle\|_{w, V_{\tilde{x}(t)}} \leq 2\varepsilon_0.$$

Varying of Trajectories in Convex Control Problems

Therefore, on the basis of the continuous dependence theorem (Theorem 4.4), we can assume that $\varepsilon_0 > 0$ is so small that the families of solutions (6.6) and (6.7) exist and belong to an arbitrary given neighborhood $V_{\bar{x}(t)}$ for all $\sigma \in \Sigma$ and for sufficiently large i (let us say, for $i \geq i_1$). The continuous dependence of $x(t; \sigma)$ and $x^{(i)}(t; \sigma)$ on $(t, \sigma) \in [t_1, t_2] \times \Sigma$ follows from Assertion 6.5.

We use the notation

$$\Delta x^{(i)}(t; \sigma) = x^{(i)}(t; \sigma) - x(t; \sigma), \quad t_1 \leq t \leq t_2, \quad \sigma \in \Sigma,$$

$$\delta v_t^{(i)}(\sigma) = \delta \mu_t^{(i)}(\sigma) - \delta \mu_t(\sigma),$$

and we write the differential equation for $\Delta x^{(i)}(t; \sigma)$:

$$\Delta \dot{x}^{(i)} = F(t, x(t; \sigma) + \Delta x^{(i)}) - F(t, x(t; \sigma)) + \langle \delta \mu_t(\sigma), f(t, x(t; \sigma) + \Delta x^{(i)}, u) \rangle$$
$$- f(t, x(t; \sigma), u) \rangle + \langle \delta v_t^{(i)}(\sigma), f(t, x^{(i)}(t; \sigma), u) \rangle$$

$$= \left\{ \int_0^1 F_x(t, x(t; \sigma) + s\Delta x^{(i)}) \, ds + \int_0^1 \langle \delta \mu_t(\sigma), f_x(t, x(t; \sigma) + s\Delta x^{(i)}, u) \rangle \, ds \right\} \Delta x^{(i)}$$

$$+ \langle \delta v_t^{(i)}(\sigma), f(t, x^{(i)}(t; \sigma), u) \rangle.$$

We denote the expression in braces by $A^{(i)}(t; \sigma)$, i.e.,

$$A^{(i)}(t; \sigma) = \int_0^1 [F_x(t, x(t; \sigma) + s\Delta x^{(i)}(t; \sigma))$$

$$+ \langle \delta \mu_t(\sigma), f_x(t, x(t; \sigma) + s\Delta x^{(i)}(t; \sigma), u) \rangle] \, ds, \quad t_1 \leq t \leq t_2, \quad \sigma \in \Sigma, \quad i \geq i_1.$$

All the measures $\delta \mu_t(\sigma)$ and $\delta \mu_t^{(i)}(\sigma)$ are concentrated on the same bounded subset of R^r. Also, we can assume that all the curves $x(t; \sigma) + s\Delta x^{(i)}(t; \sigma)$, $t_1 \leq t \leq t_2$; $0 \leq s \leq 1$, $\sigma \in \Sigma$, $i \geq i_1$, lie in the compact neighborhood $V_{\bar{x}(t)}$. Therefore, the absolute value

$$|A^{(i)}(t; \sigma)|, \quad t_1 \leq t \leq t_2, \quad \sigma \in \Sigma, \quad i \geq i_1,$$

obviously does not exceed a constant C.

From this and from Theorem 5.2, it follows that the absolute values of the fundamental matrix $\Gamma^{(i)}(t; \sigma)$, $t_1 \leq t \leq t_2$, of the equation

$$\delta \dot{x} = A^{(i)}(t; \sigma) \delta x$$

and of the matrix $G^{(i)}(t; \sigma)$ inverse to this matrix are bounded by a single constant C_1,

$$|\Gamma^{(i)}(t; \sigma)| \leq C_1, \quad |G^{(i)}(t; \sigma)| \leq C_1 \quad \forall t \in [t_1, t_2], \quad \forall \sigma \in \Sigma, \text{ and } \forall i \geq i_1.$$

We represent $\Delta x^{(i)}(t; \sigma)$ by the form

$$\Delta x^{(i)}(t; \sigma) = \Gamma^{(i)}(t; \sigma) \int_{t_1}^{t} G^{(i)}(\theta; \sigma) \langle \delta v_\theta^{(i)}(\sigma), f(\theta, x^{(i)}(\theta; \sigma), u) \rangle \, d\theta.$$

Integration by parts yields

$$\Delta x^{(i)}(t; \sigma) = \int_{t_1}^{t} \langle \delta v_\theta^{(i)}(\sigma), f(\theta, x^{(i)}(\theta; \sigma), u) \rangle \, d\theta$$

$$+ \Gamma^{(i)}(t, \sigma) \int_{t_1}^{t} G^{(i)}(\theta; \sigma) A^{(i)}(\theta; \sigma)$$

$$\times \left[\int_{t_1}^{\theta} \langle \delta v_{\theta_1}^{(i)}(\sigma), f(\theta_1, x^{(i)}(\theta_1; \sigma), u) \rangle \, d\theta_1 \right] d\theta,$$

$$\|\Delta x^{(i)}(t; \sigma)\| \leq \{1 + CC_1^2(t_2 - t_1)\} \left\| \int_{t_1}^{t} \langle \delta v_\theta^{(i)}(\sigma), f(\theta, x^{(i)}(\theta; \sigma), u) \rangle \, d\theta \right\|,$$

where

$$\left\| \int_{t_1}^{t} \langle \delta v_\theta^{(i)}(\sigma), f(\theta, x^{(i)}(\theta; \sigma), u) \rangle \, d\theta \right\| = \max_{t \in [t_1, t_2]} \left| \int_{t_1}^{t} \langle \delta v_\theta^{(i)}(\sigma), f(\theta, x^{(i)}(\theta; \sigma), u) \rangle \, d\theta \right|.$$

According to an assumption of the theorem, the sequence $\delta v_t^{(i)}(\sigma)$ converges weakly to zero as $i \to \infty$, uniformly with respect to $\sigma \in \Sigma$. Therefore, Assertion 6.2 yields

$$\| \langle \delta v_t^{(i)}(\sigma), f(t, x, u) \rangle \|_{w, V_{\tilde{x}(t)}} \to 0 \quad (i \to \infty)$$

uniformly with respect to $\sigma \in \Sigma$. From this and from Assertion 4.1, we conclude that

$$\left\| \int_{t_1}^{t} \langle \delta v_\theta^{(i)}(\sigma), f(\theta, x^{(i)}(\theta; \sigma), u) \rangle \, d\theta \right\| \to 0 \quad (i \to \infty), \tag{6.8}$$

uniformly with respect to $\sigma \in \Sigma$, for the following reasons. All the functions (6.6) satisfy a Lipschitz condition with the same constant L which does not

Varying of Trajectories in Convex Control Problems

depend on $\sigma \in \Sigma$ and $i \geqslant i_1$ as follows:

$$\begin{aligned}
\left|x^{(i)}(t'; \sigma) - x^{(i)}(t''; \sigma)\right| &\leqslant \left|\int_{t'}^{t''} \left|F(t, x^{(i)}(t; \sigma))\right| dt\right| \\
&\quad + \left|\int_{t'}^{t''} \left|\langle \delta\mu_t^{(i)}(\sigma), f(t, x^{(i)}(t; \sigma), u)\rangle\right| dt\right| \\
&\leqslant \left|\int_{t'}^{t''} \sup_{(t,x) \in V_{\tilde{x}(t)}} \left\{|F(t, x)| + |\langle \delta\mu_t^{(i)}(\sigma), f(t, x, u)\rangle|\right\} dt\right| \\
&\leqslant L|t' - t''|.
\end{aligned}$$

Thus, these functions form an equicontinuous family.

From (6.8) it follows that

$$\|\Delta x^{(i)}(t; \sigma)\| \to 0 \qquad (i \to \infty),$$

uniformly with respect to $\sigma \in \Sigma$. This concludes the proof of the theorem.

7

Proof of the Maximum Principle

In this chapter, we prove the maximum principle for the convex optimal problem in the so-called *integral form*. The maximum principle formulated in Chapter 1 follows immediately from this form.

As we shall show here, the integral form of the maximum principle is equivalent to the maximum principle "in Pontryagin's form" for the classes of controls Ω_U and $\mathfrak{M}_U$ which we consider (see below). However, there is no such equivalence for more general classes of controls, and the integral form turns out to be more universal.

7.1. The Integral Maximum Condition, the Pontryagin Maximum Condition, and Their Equivalence

Let

$$H(t, x, \psi, u) = \psi f(t, x, u)$$

be the Hamiltonian of the optimal problem (1.3), and let

$$\dot{x} = \frac{\partial H}{\partial \psi} = f(t, x, u), \qquad \dot{\psi} = -\frac{\partial H}{\partial x} = -\psi f_x(t, x, u)$$

be the corresponding canonical system with respect to x and ψ, and containing the parameter u.

In Chapter 1, we substituted for u admissible controls $u(t) \in \Omega_U$. Now, we shall substitute arbitrary generalized controls $\mu_t \in \mathfrak{M}_U$:

$$\dot{x} = \frac{\partial H}{\partial \psi} = \langle \mu_t, f(t, x, u) \rangle,$$

$$\dot{\psi} = -\frac{\partial H}{\partial x} = -\psi \langle \mu_t, f_x(t, x, u) \rangle. \quad (7.1)$$

We obtain the previous system when $\mu_t = \delta_{u(t)}$.

The generalized control μ_t is not determined uniquely by the canonical system (7.1). In order to solve this system in a unique way with given initial values of x and ψ, we need, as before, an additional condition which "eliminates" μ_t. We shall formulate this condition in two equivalent forms, namely, in the form of the *integral maximum condition* and the *Pontryagin maximum condition*.

Let $x(t)$ and $\psi(t)$ be arbitrary continuous functions on $[t_1, t_2]$. We shall say that a generalized control $\hat{\mu}_t \in \mathfrak{M}_U$ satisfies the *integral maximum condition along* $x(t), \psi(t), t_1 \leq t \leq t_2$, if

$$\int_{t_1}^{t_2} \langle \hat{\mu}_t, H(t, x(t), \psi(t), u) \rangle \, dt = \sup_{\mu_t \in \mathfrak{M}_U} \int_{t_1}^{t_2} \langle \mu_t, H(t, x(t), \psi(t), u) \rangle \, dt. \quad (7.2)$$

We shall also say that this control satisfies the *Pontryagin maximum condition* if, for almost all $t \in [t_1, t_2]$,

$$\langle \hat{\mu}_t, H(t, x(t), \psi(t), u) \rangle = \sup_\mu \langle \mu, H(t, x(t), \psi(t), u) \rangle,$$

where the supremum is taken over all probability measures μ concentrated on $U \subset R^r$.

We shall show that, for any $t \in [t_1, t_2]$,

$$\sup_\mu \langle \mu, H(t, x(t), \psi(t), u) \rangle = \sup_{u \in U} H(t, x(t), \psi(t), u) = M(t, x(t), \psi(t)),$$

so that the Pontryagin maximum condition can be written in the form

$$\langle \hat{\mu}_t, H(t, x(t), \psi(t), u) \rangle = \sup_\mu \langle \mu, H(t, x(t), \psi(t), u) \rangle = M(t, x(t), \psi(t)) \quad (7.3)$$

for almost all $t \in [t_1, t_2]$.

The inequality

$$\sup_\mu \langle \mu, H(t, x(t), \psi(t), u) \rangle = \sup_\mu \psi(t) \langle \mu, f(t, x(t), u) \rangle \geq M(t, x(t), \psi(t))$$

is obvious. Indeed, it is enough to take δ_u for μ, where u is any point of U.

In order to prove the opposite inequality, we note that any probability measure μ which is concentrated on U satisfies (see Assertion 2.2) the condition

$$\langle \mu, f(t, x(t), u)\rangle \in \text{conv}\{f(t, x(t), u): u \in U\} = \text{conv}\, P(t, x(t)).$$

Therefore,

$$\psi(t)\langle \mu, f(t, x(t), u)\rangle \leq \sup_{p \in \text{conv}\, P(t,x(t))} \psi(t)p.$$

However, since for any n-dimensional row ψ and any subset $P \subset R^n$, we have

$$\sup_{p \in \text{conv}\,\bar{P}} \psi p = \sup_{p \in P} \psi p *$$

($\bar{P}$ is the closure of P), we obtain the relations

$$\psi(t)\langle \mu, f(t, x(t), u)\rangle \leq \sup_{p \in P(t,x(t))} \psi(t)p = \sup_{u \in U} \psi(t) f(t, x(t), u)$$

and, therefore, the formula (7.3).

Assertion 7.1. The maximum conditions in the integral form (7.2) and in the Pontryagin form (7.3) are equivalent.

Proof. In order to prove the implication (7.2) $\Rightarrow$ (7.3), we assume that $\theta \in (t_1, t_2)$ is an arbitrary Lebesgue point of the function

$$F(t) = \langle \hat{\mu}_t, H(t, x(t), \psi(t), u)\rangle, \quad t_1 \leq t \leq t_2.$$

For any small $h > 0$, we define the generalized control $\mu_t^h \in \mathfrak{M}_U$ by the formula

$$\mu_t^h = \hat{\mu}_t \quad \text{for } t_1 \leq t \leq \theta \text{ or } \theta + h \leq t \leq t_2,$$
$$\mu_t^h = \mu \quad \text{for } \theta < t < \theta + h,$$

where μ is an arbitrary probability measure concentrated on U. It follows

---

*Since

$$\sup_{p \in P} \psi p \leq \sup_{p \in \text{conv}\,\bar{P}} \psi p,$$

we can assume that the left-hand side of the inequality is finite. Assume that it is equal to s. Then the closed convex set of all p that satisfy the condition $\psi p \leq s$ ($\psi \neq 0$) contains the set P and, therefore, also conv $\bar{P}$. Thus, we obtain the opposite inequality

$$\sup_{p \in \text{conv}\,\bar{P}} \psi p \leq s = \sup_{p \in P} \psi p.$$

from (7.2) that

$$\frac{1}{h}\int_{\theta}^{\theta+h} \langle \hat{\mu}_t, H(t, x(t), \psi(t), u)\rangle \, dt \geq \frac{1}{h}\int_{\theta}^{\theta+h} \langle \mu, H(t, x(t), \psi(t), u)\rangle \, dt.$$

The point θ is a Lebesgue point of both integrands (the right-hand side function is continuous in t). Therefore, in the limit as $h \to 0$, we obtain the inequality

$$\langle \hat{\mu}_\theta, H(\theta, x(\theta), \psi(\theta), u)\rangle \geq \langle \mu, H(\theta, x(\theta), \psi(\theta), u)\rangle$$

for all Lebesgue points $\theta \in [t_1, t_2]$. This inequality is equivalent to the condition (7.3), since the Lebesgue points form a set of full (Lebesgue) measure in $[t_1, t_2]$.

The implication (7.3) $\Rightarrow$ (7.2) is obtained very simply, namely, by the integration from t_1 up to t_2 of the following relations, which hold for almost all $t \in [t_1, t_2]$:

$$\langle \hat{\mu}_t, H(t, x(t), \psi(t), u)\rangle = \sup_{\mu} \langle \mu, H(t, x(t), \psi(t), u)\rangle \geq \langle \mu_t, H(t, x(t), \psi(t), u)\rangle,$$

where $\mu_t \in \mathfrak{M}_U$ is an arbitrary generalized control.

If we make the substitution

$$\hat{\mu}_t = \delta_{\hat{u}(t)}, \qquad \hat{u}(t) \in \Omega_U,$$

in the formula (7.3), then it becomes the maximum condition (1.9).

7.2. The Maximum Principle in the Class of Generalized Controls

We consider the convex optimal problem

$$\dot{x} = \langle \mu_t, f(t, x, u)\rangle, \qquad \mu_t \in \mathfrak{M}_U,$$
$$t = t_1, \qquad x(t_1) = x_1, \qquad x(t_2) = x_2, \qquad (7.4)$$
$$t_2 - t_1 \to \min.$$

By analogy with the definition of an extremal given in Chapter 1, we shall say that any solution

$$\tilde{\mu}_t, \tilde{x}(t), \tilde{\psi}(t), \qquad t_1 \leq t \leq t_2,$$

of the canonical system (7.1) with the boundary conditions

$$t = t_1, \qquad \tilde{x}(t_1) = x_1, \qquad \tilde{x}(t_2) = x_2$$

Proof of the Maximum Principle

is an *extremal of the convex optimal problem* (7.4), if $\tilde{\psi}(t) \neq 0$ and if $\tilde{\mu}_t$ satisfies the maximum condition (7.2) or (7.3) along $\tilde{x}(t)$, $\tilde{\psi}(t)$, $t_1 \leqslant t \leqslant t_2$.

If a generalized control is also an ordinary control, $\tilde{\mu}_t = \delta_{\tilde{u}(t)}$, then an extremal of the convex problem

$$\delta_{\tilde{u}(t)}, \quad \tilde{x}(t), \quad \tilde{\psi}(t)$$

is at the same time an extremal of the initial optimal problem (1.3), in accordance with the definition given before.

Continuing this analogy, one can attempt to formulate the maximum principle for the convex optimal problem (7.4) in the following way: If $\tilde{\mu}_t$, $\tilde{x}(t)$, $\tilde{\psi}(t)$ is a solution of the convex optimal problem, then there exists an absolutely continuous function $\tilde{\psi}(t) \neq 0$ such that the system of functions

$$\tilde{\mu}_t, \quad \tilde{x}(t), \quad \tilde{\psi}(t)$$

is an extremal of this problem.

However, it is easy to see that, although this assertion is true, it is not sufficiently strong in order to imply the maximum principle formulated in Chapter 1. Indeed, for $\tilde{\mu}_t = \delta_{\tilde{u}(t)} \in \Omega_U$ this formulation can guarantee the existence of $\tilde{\psi}(t)$ with the properties that we need only if the control $\delta_{\tilde{u}(t)}$ is optimal in the class of all generalized controls $\mathfrak{M}_U$, whereas the maximum principle formulated in Chapter 1 asserts that a required $\tilde{\psi}(t)$ exists even in the case where $\delta_{\tilde{u}(t)}$ is optimal in the narrower class Ω_U. For this reason, we shall formulate and prove a stronger necessary condition for optimality in the convex problem (7.4). The maximum principle formulated in Chapter 1 follows immediately from this condition.

We shall say that a generalized control $\tilde{\mu}_t$ and the corresponding trajectory $\tilde{x}(t)$, $t_1 \leqslant t \leqslant t_2$, of the equation

$$\dot{x} = \langle \tilde{\mu}_t, f(t, x, u) \rangle$$

with the boundary conditions

$$t = t_1, \quad \tilde{x}(t_1) = x_1, \quad \tilde{x}(t_2) = x_2$$

yield a *solution of problem* (7.4) *optimal in the weakened sense* if there is no admissible control $\delta_{u(t)} \in \Omega_U$ which transfers the phase point along a trajectory of the equation

$$\dot{x} = \langle \delta_{u(t)}, f(t, x, u) \rangle = f(t, x, u(t))$$

with the same boundary conditions during a time shorter than $t_2 - t_1$.

Obviously, any solution of the convex optimal problem (7.4) is also optimal in the weakened sense. Therefore, the concept of "weakened optimality" is indeed weaker than the concept of optimality. Obviously, both concepts of optimality coincide in the class of ordinary controls.

We can now formulate the maximum principle for the convex optimal problem.

Theorem 7.1. (*The Maximum Principle for the Convex Optimal Problem*). Let a generalized control $\bar{\mu}_t$ and the corresponding trajectory $\tilde{x}(t)$, $t_1 \leq t \leq t_2$, of the equation

$$\dot{x} = \langle \bar{\mu}_t, f(t, x, u) \rangle$$

with the boundary conditions

$$t = t_1, \qquad \tilde{x}(t_1) = x_1, \qquad \tilde{x}(t_2) = x_2$$

be optimal in the weakened sense in the problem (7.4). Then there exists a nonzero absolutely continuous function $\tilde{\psi}(t)$ such that

$$\bar{\mu}_t, \tilde{x}(t), \tilde{\psi}(t)$$

is a solution of the canonical system (7.1) and $\bar{\mu}_t$ satisfies the integral maximum condition (7.2) along $\tilde{x}(t)$, $\tilde{\psi}(t)$. In other words, the system of functions $\bar{\mu}_t, \tilde{x}(t), \tilde{\psi}(t)$ is an extremal of the convex optimal problem (7.4).

Moreover, the function of t

$$M(t, \tilde{x}(t), \tilde{\psi}(t)) = \sup_{u \in U} H(t, \tilde{x}(t), \tilde{\psi}(t), u)$$

is continuous on the entire interval $t_1 \leq t \leq t_2$ and

$$M(t_2, \tilde{x}(t_2), \tilde{\psi}(t_2)) \geq 0.$$

7.3. Construction of the Cone of Variations

Before beginning to prove Theorem 7.1, we shall construct the so-called cone of first-order variations and prove its basic property.

Let $\bar{\mu}_t$ be an arbitrary (not necessarily optimal) generalized control, and let $\tilde{x}(t)$ with $t_1 \leq t \leq t_2$ be the corresponding trajectory of the equation

$$\dot{x} = \langle \bar{\mu}_t, f(t, x, u) \rangle = F(t, x) \qquad (7.5)$$

with the boundary conditions

$$t = t_1, \qquad x(t) = x_1, \qquad x(t_2) = t_2.$$

Proof of the Maximum Principle

As in Chapter 6, we shall use the notation

$$\delta x(t; \delta\mu_\theta), \quad t_1 \leq t \leq t_2, \quad \delta\mu_t \in \delta\mathfrak{M}_{\bar{\mu}_t},$$

to represent the solution of the equation of variations for (7.5) along $\tilde{x}(t)$:

$$\delta\dot{x} = F_x(t, \tilde{x}(t))\delta x + \langle \delta\mu_t, f(t, \tilde{x}(t), u)\rangle$$

with zero initial condition

$$\delta x(t_1; \delta\mu_\theta) = 0.$$

The function $\tilde{x}(t)$ can be nondifferentiable at the point t_2. Nevertheless, for an appropriate sequence of numbers η_j converging to zero, there always exists a *finite* limit

$$\lim_{j\to\infty} \frac{\tilde{x}(t_2) - \tilde{x}(t_2 - \eta_j)}{\eta_j}, \quad \eta_j > 0.$$

This is so because the unit measure $\bar{\mu}_t$ is concentrated on the same bounded set for all t, and hence

$$|\tilde{x}(t_2) - \tilde{x}(t_2 - \eta)| \leq \int_{t_2-\eta}^{t_2} |\dot{\tilde{x}}(t)|\, dt = \int_{t_2-\eta}^{t_2} |\langle \bar{\mu}_t, f(t, \tilde{x}(t), u)\rangle|\, dt \leq \eta \cdot \text{const}.$$

We choose any one of these limits and denote it by F. With the aid of the vector F, we construct in R^n the set

$$Q = \{\gamma\delta x(t_2; \delta\mu_\theta) - \vartheta F : \delta\mu_t \in \delta\mathfrak{M}_{\bar{\mu}_t}, \gamma \geq 0, \vartheta \geq 0\}. \tag{7.6}$$

It is easy to see that Q is a *convex cone* in R^n with vertex at the origin, i.e., $\delta x', \delta x'' \in Q$ and $\alpha, \beta \geq 0$ imply

$$\alpha\delta x' + \beta\delta x'' \in Q.$$

Indeed, since $\delta\mathfrak{M}_{\bar{\mu}_t}$ is a convex set, we have

$$\alpha(\gamma'\delta x(t_2; \delta\mu'_\theta) - \vartheta' F) + \beta(\gamma''\delta x(t; \delta\mu''_\theta) - \vartheta'' F)$$

$$= \delta x(t_2; \alpha\gamma'\delta\mu'_\theta + \beta\gamma''\delta\mu''_\theta) - (\alpha\vartheta' + \beta\vartheta'')F$$

$$= \frac{1}{\varepsilon}\delta x(t_2; \varepsilon\alpha\gamma'\delta\mu'_\theta + \varepsilon\beta\gamma''\delta\mu''_\theta) - (\alpha\vartheta' + \beta\vartheta'')F, \quad \varepsilon > 0,$$

and the assertion follows from the fact that

$$\varepsilon\alpha\gamma'\delta\mu'_t + \varepsilon\beta\gamma''\delta\mu''_t \in \delta\mathfrak{M}_{\bar{\mu}_t}$$

for $\varepsilon \geq 0$ sufficiently small.

The set Q can be also described in the following way: It is the cone in R^n with vertex at the origin subtended by the convex hull of the set of points

$$\{\delta x(t_2; \delta\mu_\theta), -F : \delta\mu_t \in \delta\mathfrak{M}_{\tilde{\mu}_t}\} \subset R^n,$$

i.e., by the set of points of the form

$$\alpha\delta x(t_2; \delta\mu_\theta) - \beta F, \qquad \delta\mu_t \in \delta\mathfrak{M}_{\tilde{\mu}_t}, \qquad \alpha \geq 0, \quad \beta \geq 0, \quad \alpha + \beta = 1.$$

The cone Q is called the *cone of first-order variations* of the convex control problem

$$\dot{x} = \langle \mu_t, f(t, x, u) \rangle, \qquad \mu_t \in \mathfrak{M}_U,$$

constructed for the point $\tilde{x}(t_2)$ of the trajectory $\tilde{x}(t)$, $t_1 \leq t \leq t_2$.

If the function $\tilde{x}(t)$ is differentiable at the point t_2, then we have the unique vector of the form indicated,

$$F = \frac{d\tilde{x}(t_2)}{dt},$$

and there is no arbitrariness in the construction of the cone Q. This is so, e.g., if the point t_2 is a Lebesgue point of the function $F(t, \tilde{x}(t))$. However, in the general case Q depends on the choice of F.

The vertex of the cone of variations Q (the origin of R^n) can either belong to the boundary of the set $Q \subset R^n$ or an interior point. In the latter case, the set Q, obviously, coincides with R^n.

The following important assertion expresses the basic property of the cone of variations.

Assertion 7.2. If the cone (7.6) coincides with R^n, then any point of a sufficiently small neighborhood of $\tilde{x}(t_2)$ can be reached with the aid of a piecewise-continuous admissible control $\delta_{u(t)} \in \Omega_U$ in a time less than $t_2 - t_1$, if the motion starts at the point x_1 and at the instant of time $t = t_1$. This means that the equation

$$\dot{x} = \langle \delta_{u(t)}, f(t, x, u) \rangle = f(t, x, u(t))$$

has the solution

$$x(t), \qquad t_1 \leq t \leq \hat{t}_2, \quad \hat{t}_2 < t_2,$$

with the boundary conditions

$$x(t_1) = \tilde{x}(t_1), \qquad x(\hat{t}_2) = \tilde{x}(t_2).$$

Proof of the Maximum Principle

The proof presented below reduces this assertion to the theorem on the existence of a fixed point under a continuous mapping of an n-dimensional simplex into itself.

Proof. In $Q = R^n$ we choose $1+n$ points

$$\delta x^{(0)}, \ldots, \delta x^{(n)}$$

which form vertices of an n-dimensional simplex containing the origin as an interior point. We denote this simplex by X:

$$X = \left\{ \delta x : \delta x = \sum_{k=0}^{n} \lambda_k \delta x^{(k)}, \lambda_k \geq 0, \sum_{k=0}^{n} \lambda_k = 1 \right\}.$$

According to (7.6), we can choose perturbations and nonnegative numbers

$$\delta \mu_t^{(k)} \in \delta \mathfrak{M}_{\bar{\mu}_t}, \quad \gamma^{(k)} \geq 0, \quad \vartheta^{(k)} \geq 0, \quad k=0, 1, \ldots, n,$$

such that

$$\delta x^{(k)} = \delta x(t_2; \gamma^{(k)} \delta \mu_\theta^{(k)}) - \vartheta^{(k)} F. \tag{7.7}$$

It is necessary for the proof of the assertion we are giving here, to assume that all $\vartheta^{(k)} > 0$, $k = 0, 1, \ldots, n$. We can do this without loss of generality for the following reason: The points $\delta x^{(k)}$, $k = 0, 1, \ldots, n$, as vertices of an n-dimensional simplex in R^n, are in general position in R^n. Therefore, all $\vartheta^{(k)} \geq 0$ can be increased to certain $\hat{\vartheta}^{(k)} > 0$ and the increments can be so small that the slightly changed points

$$\delta \hat{x}^{(k)} = \delta x(t_2; \gamma^{(k)} \delta \mu_\theta^{(k)}) - \hat{\vartheta}^{(k)} F, \quad k=0, 1, \ldots, n,$$

are also in general position in R^n, so that the simplex spanned by them contains the origin as an interior point.

Since in all subsequent constructions we make use of the values $\delta \mu_t^{(k)}$ only for $t \in [t_1, t_2]$, we assume in addition that $\delta \mu_t^{(k)} = 0$ for $t \notin [t_1, t_2]$. This is done for the convenience of referring to the assertions proved in Chapter 6.

We use the notation $d > 0$ to represent the distance from the origin to the boundary of the simplex X. We assume that the points $\vartheta^{(k)}$ are numbered in the order in which they increase:

$$0 < \vartheta^{(0)} \leq \vartheta^{(1)} \leq \cdots \leq \vartheta^{(n)}. \tag{7.8}$$

We use the notation Λ to represent the n-dimensional simplex

$$\Lambda = \left\{ \lambda = (\lambda_0, \ldots, \lambda_n) : \lambda_k \geq 0, \sum_{k=0}^{n} \lambda_k = 1 \right\}.$$

According to Theorem 6.1, for all sufficiently small ε (let us say, for ε with $0 \leq \varepsilon \leq \varepsilon_1$, $\varepsilon_1 > 0$), the family of perturbations

$$\delta\mu_t(\varepsilon, \lambda) = \varepsilon \sum_{k=0}^{n} \lambda_k \gamma^{(k)} \delta\mu_t^{(k)}, \qquad \lambda = (\lambda_0, \ldots, \lambda_n) \in \Lambda,$$

generates the family of solutions

$$x(t; \varepsilon, \lambda) = \tilde{x}(t) + \varepsilon \sum_{k=0}^{n} \lambda_k \delta x(t; \gamma^{(k)} \delta\mu_\theta^{(k)}) + \Delta_2 x(t; \varepsilon, \lambda), \qquad t_1 \leq t \leq t_2, \quad (7.9)$$

$$x(t_1; \varepsilon, \lambda) = x_1,$$

of the perturbed equation

$$\dot{x} = F(t, x) + \langle \delta\mu_t(\varepsilon, \lambda), f(t, x, u) \rangle.$$

Here, $\Delta_2 x(t; \varepsilon, \lambda)$ satisfies the estimate

$$\|\Delta_2 x(t; \varepsilon, \lambda)\| \leq o(\varepsilon) \qquad \forall \lambda \in \Lambda, \quad \frac{o(\varepsilon)}{\varepsilon} \to 0 \qquad (\varepsilon \to 0),$$

and the function $x(t; \varepsilon, \lambda)$ depends continuously on

$$(t, \varepsilon, \lambda) \in [t_1, t_2] \times [0, \varepsilon_1] \times \Lambda.$$

We now find positive continuous functions

$$\varepsilon(\lambda), \vartheta(\lambda), \qquad \lambda \in \Lambda,$$

that satisfy the condition

$$0 < \varepsilon(\lambda)\vartheta(\lambda) \leq t_2 - t_1, \qquad \varepsilon(\lambda) \leq \varepsilon_1 \qquad \forall \lambda \in \Lambda,$$

and are such that the continuous mapping

$$S: \Lambda \to R^n$$

defined by the formula

$$S(\lambda) = x(t_2 - \varepsilon(\lambda)\vartheta(\lambda); \varepsilon(\lambda), \lambda) - \tilde{x}(t_2), \qquad \lambda \in \Lambda,$$

essentially covers a neighborhood $\hat{V}$ of the origin of R^n. This means that there exists a positive number δ such that, for any continuous function $\omega(\lambda)$, $\lambda \in \Lambda$, which takes on values in R^n and satisfies the inequality

$$|\omega(\lambda)| \leq \delta \qquad \forall \lambda \in \Lambda,$$

the image of the simplex Λ under the mapping

$$\lambda \to S(\lambda) + \omega(\lambda)$$

Proof of the Maximum Principle

will cover the entire neighborhood $\hat{V}$. In other words, the following equation for λ,

$$S(\lambda) + \omega(\lambda) = \delta x, \quad \lambda \in \Lambda,$$

is solvable for any $\delta x \in \hat{V}$.

We obtain as a corollary for $\omega(\lambda) \equiv 0$ that any point of the neighborhood

$$\tilde{x}(t_2) + \hat{V} = \{x : x = \tilde{x}(t_2) + \delta x, \delta x \in \hat{V}\}$$

can be reached along trajectories of the family (7.9) at some time $t_2 - \varepsilon(\lambda)\vartheta(\lambda) < t_2$.

We define

$$\vartheta(\lambda) = \sum_{k=0}^{n} \lambda_k \vartheta^{(k)}, \quad \lambda = (\lambda_0, \ldots, \lambda_n) \in \Lambda,$$

where $\vartheta^{(k)}$ are given in (7.7). When λ ranges over Λ, the value of $\vartheta(\lambda)$ is contained between $\vartheta^{(0)}$ and $\vartheta^{(n)}$, because in accordance with (7.8), we have

$$0 < \vartheta^{(0)} \leq \vartheta^{(0)} + \sum_{k=1}^{n} \lambda_k(\vartheta^{(k)} - \vartheta^{(0)}) = \vartheta(\lambda) = \vartheta^{(n)} + \sum_{k=1}^{n} \lambda_k(\vartheta^{(k)} - \vartheta^{(n)}) \leq \vartheta^{(n)}.$$

Let a sequence of numbers $\eta_j > 0, j = 1, 2, \ldots$, satisfy the condition

$$F = \frac{\tilde{x}(t_2) - \tilde{x}(t_2 - \eta_j)}{\eta_j} + v_j, \quad \eta_j \to 0, \quad v_j \to 0 \quad (j \to \infty). \tag{7.10}$$

The positive continuous functions

$$\varepsilon_j(\lambda) = \frac{\eta_j}{\vartheta(\lambda)}, \quad \lambda \in \Lambda,$$

satisfy the inequalities

$$\frac{\eta_j}{\vartheta^{(n)}} \leq \varepsilon_j(\lambda) \leq \frac{\eta_j}{\vartheta^{(0)}} \quad \forall \lambda \in \Lambda, \tag{7.11}$$

and, for large j,

$$\varepsilon_j(\lambda) \leq \varepsilon_1 \quad \forall \lambda \in \Lambda.$$

The following functions are defined for these j:

$$z_j(\lambda) = x(t_2 - \varepsilon_j(\lambda)\vartheta(\lambda); \varepsilon_j(\lambda), \lambda) - \tilde{x}(t_2) = x(t_2 - \eta_j; \varepsilon_j(\lambda), \lambda) - \tilde{x}(t_2), \quad \lambda \in \Lambda.$$

On the basis of (7.10), these functions can be written in the form

$$z_j(\lambda) = x(t_2 - \eta_j; \varepsilon_j(\lambda), \lambda) - \tilde{x}(t_2 - \eta_j) - \eta_j F + \eta_j v_j.$$

We express the difference
$$x(t_2-\eta_j;\varepsilon_j(\lambda),\lambda)-\tilde{x}(t_2-\eta_j)$$
with the aid of formula (7.9), substituting in this formula the following value:
$$t=t_2-\eta_j=t_2-\varepsilon_j(\lambda)\vartheta(\lambda)=t_2-\varepsilon_j(\lambda)\sum_{k=0}^{n}\lambda_k\vartheta^{(k)}.$$
Then we obtain [see (7.7)]
$$z_j(\lambda)=\varepsilon_j(\lambda)\left\{\sum_{k=0}^{n}\lambda_k[\delta x(t_2;\gamma^{(k)}\delta\mu_\theta^{(k)})-\vartheta^{(k)}F]\right.$$
$$+\sum_{k=0}^{n}\lambda_k[\delta x(t_2-\eta_j;\gamma^{(k)}\delta\mu_\theta^{(k)})-\delta x(t_2;\gamma^{(k)}\delta\mu_\theta^{(k)})]$$
$$\left.+\frac{\Delta_2 x(t_2-\eta_j;\varepsilon_j(\lambda),\lambda)}{\varepsilon_j(\lambda)}+\vartheta(\lambda)v_j\right\}$$
$$=\varepsilon_j(\lambda)\left\{\sum_{k=0}^{n}\lambda_k\delta x^{(k)}+R_j(\lambda)\right\},$$
where
$$R_j(\lambda)=\sum_{k=0}^{n}\lambda_k[\delta x(t_2-\eta_j;\gamma^{(k)}\delta\mu_\theta^{(k)})-\delta x(t_2;\gamma^{(k)}\delta\mu_\theta^{(k)})]$$
$$+\frac{\Delta_2 x(t_2-\eta_j;\varepsilon_j(\lambda),\lambda)}{\varepsilon_j(\lambda)}+\vartheta(\lambda)v_j.$$
As $j\to\infty$, we have
$$R_j(\lambda)\to 0$$
uniformly with respect to $\lambda\in\Lambda$.

We fix an index $j=\hat{j}$ for which
$$|R_{\hat{j}}(\lambda)|\leq\frac{d}{3}\quad\forall\lambda\in\Lambda,$$
and use the notation
$$\varepsilon(\lambda)=\varepsilon_{\hat{j}}(\lambda),\quad R(\lambda)=R_{\hat{j}}(\lambda),$$
$$S(\lambda)=z_{\hat{j}}(\lambda)=x(t_2-\varepsilon(\lambda)\vartheta(\lambda);\varepsilon(\lambda),\lambda)-\tilde{x}(t_2)=\varepsilon(\lambda)\left\{\sum_{k=0}^{n}\lambda_k\delta x^{(k)}+R(\lambda)\right\},\quad\lambda\in\Lambda.$$

We shall prove that, if
$$\hat{V}=\left\{\delta x;|\delta x|\leq\frac{\eta_{\hat{j}}}{\vartheta^{(n)}}\frac{d}{3}\right\},$$

Proof of the Maximum Principle

and if $\omega(\lambda)$ is an arbitrary continuous function that satisfies the inequality

$$|\omega(\lambda)| \leq \frac{\eta_{\hat{j}}}{\vartheta^{(n)}} \frac{d}{3},$$

then the equation for λ

$$S(\lambda) + \omega(\lambda) = \delta x, \qquad \lambda \in \Lambda, \tag{7.12}$$

is solvable for any $\delta x \in \hat{V}$. We rewrite this equation in the form

$$\sum_{k=0}^{n} \lambda_k \delta x^{(k)} = \frac{\delta x - \omega(\lambda)}{\varepsilon(\lambda)} - R(\lambda), \qquad \lambda \in \Lambda.$$

The left-hand side can be considered as the nonsingular affine mapping

$$L: \Lambda \to X$$

of the n-dimensional simplex Λ onto the n-dimensional simplex X which acts according to the rule

$$\lambda = (\lambda_0, \ldots, \lambda_n) \mapsto \sum_{k=0}^{n} \lambda_k \delta x^{(k)}.$$

Therefore, the equation can be written in the form

$$L(\lambda) = \frac{\delta x - \omega(\lambda)}{\varepsilon(\lambda)} - R(\lambda), \qquad \lambda \in \Lambda.$$

The vector on the right-hand side belongs to the simplex X for all $\lambda \in \Lambda$, because [see the inequalities (7.11)]

$$\left| \frac{\delta x - \omega(\lambda)}{\varepsilon(\lambda)} - R(\lambda) \right| \leq \frac{1}{\varepsilon(\lambda)} \frac{\eta_{\hat{j}}}{\vartheta^{(n)}} \frac{2}{3} d + \frac{d}{3} \leq d.$$

Therefore, by denoting

$$L^{-1}: X \to \Lambda$$

as the mapping inverse to L, we can rewrite the last equation in the equivalent form

$$\lambda = L^{-1}\left(\frac{\delta x - \omega(\lambda)}{\varepsilon(\lambda)} - R(\lambda) \right), \qquad \lambda \in \Lambda.$$

Every fixed point of the continuous mapping of the simplex Λ into itself given by the formula

$$\lambda \mapsto L^{-1}\left(\frac{\delta x - \omega(\lambda)}{\varepsilon(\lambda)} - R(\lambda)\right)$$

is a solution of the equation thus obtained. Such a point always exists by the well-known Brouwer theorem.

Thus, equation (7.12) has at least one solution, and every solution $\hat{\lambda}$ of this equation yields

$$x(t_2 - \varepsilon(\hat{\lambda})\vartheta(\hat{\lambda}); \varepsilon(\hat{\lambda}), \hat{\lambda}) = \tilde{x}(t_2) + \delta x.$$

The generalized control

$$\tilde{\mu}_t + \varepsilon(\lambda) \sum_{k=0}^{n} \hat{\lambda}_k \delta\mu_t^{(k)}$$

to which the trajectory on the left-hand side corresponds is not, generally speaking, an ordinary (admissible) control, much less a piecewise-constant control. Therefore, the equality just obtained does not yet prove Assertion 7.2. In order to finish the proof, we should turn at this point to the approximation lemma (Theorem 3.2), which asserts that the family of generalized controls

$$\tilde{\mu}_t + \delta\mu_t(\varepsilon, \lambda), \qquad (\varepsilon, \lambda) \in [0, \varepsilon_1] \times \Lambda,$$

can be approximated with arbitrary accuracy in the sense of weak convergence by piecewise-constant admissible controls.

Here is the precise statement of this lemma in the form that we need:
There exists a sequence of piecewise-constant admissible controls

$$u^{(i)}(t; \varepsilon, \lambda) \in \Omega_U, \qquad i = 1, 2, \ldots,$$

that depend strongly continuously on the parameter

$$\sigma = (\varepsilon, \lambda) \in [0, \varepsilon_1] \times \Lambda,$$

and are such that: (i) the measures $\delta_{u^{(i)}}(t; \varepsilon, \lambda)$ are concentrated on a single bounded subset of the set $U \subset R^r$ and (ii) the sequence of perturbations

$$\delta v_t^{(i)}(\varepsilon, \lambda) = \delta_{u^{(i)}(t, \varepsilon, \lambda)} - (\tilde{\mu}_t + \delta\mu_t(\varepsilon, \lambda)) \in \delta\mathfrak{M}_{\tilde{\mu}_t + \delta\mu_t(\varepsilon, \lambda)}$$

converges weakly to zero as $i \to \infty$, uniformly with respect to $(\varepsilon, \lambda) \in [0, \varepsilon_1] \times \Lambda$.

Proof of the Maximum Principle

Thus, the sequence of perturbations

$$\delta\mu_t(\varepsilon, \lambda) + \delta v_t^{(i)}(\varepsilon, \lambda) \in \delta\mathfrak{M}_{\bar\mu_t}$$

converges weakly to the perturbation $\delta\mu_t(\varepsilon, \lambda)$ as $i \to \infty$, uniformly with respect to $(\varepsilon, \lambda) \in [0, \varepsilon_1] \times \Lambda$. It then follows from Theorem 6.2 that for all sufficiently small $\varepsilon \geqslant 0$ and large i, let us say, for

$$(\varepsilon, \lambda) \in [0, \varepsilon_1] \times \Lambda, \qquad i \geqslant i_1,$$

there exists a family of solutions

$$y^{(i)}(t; \varepsilon, \lambda) = y(t; \delta_{u^{(i)}(t,\varepsilon,\lambda)}), \qquad t_1 \leqslant t \leqslant t_2,$$
$$y^{(i)}(t_1; \varepsilon, \lambda) = x_1$$

of the equation

$$\dot{x} = \langle \delta_{u^{i}(t,\varepsilon,\lambda)}, f(t, x, u) \rangle = F(t, x) + \langle \delta\mu_t(\varepsilon, \lambda), f(t, x, u) \rangle + \langle \delta v_t^{(i)}(\varepsilon, \lambda), f(t, x, u) \rangle,$$

which depends continuously on $(t, \varepsilon, \lambda) \in [t_1, t_2] \times [0, \varepsilon_1] \times \Lambda$ and satisfies the condition

$$\|y^{(i)}(t; \varepsilon, \lambda) - x(t; \varepsilon, \lambda)\| \to 0 \qquad (i \to \infty).$$

Here, $x(t; \varepsilon, \lambda)$ is the family (7.9), and the convergence to zero is uniform with respect to $(\varepsilon, \lambda) \in [0, \varepsilon_1] \times \Lambda$.

We use the notation

$$y^{(i)}(t_2 - \varepsilon(\lambda)\vartheta(\lambda); \varepsilon(\lambda), \lambda) - x(t_2 - \varepsilon(\lambda)\vartheta(\lambda); \varepsilon(\lambda), \lambda) = w^{(i)}(\lambda).$$

It follows from what we have said that, for all i sufficiently large,

$$|w^{(i)}(\lambda)| \leqslant \frac{\eta \hat{j}}{\vartheta^{(n)}} \frac{d}{3}.$$

Therefore, the equation for λ

$$y^{(i)}(t_2 - \varepsilon(\lambda)\vartheta(\lambda); \varepsilon(\lambda), \lambda) - \tilde{x}(t_2) = S(\lambda) + w^{(i)}(\lambda) = \delta x, \qquad \lambda \in \Lambda.$$

is solvable for any $\delta x \in \hat{V}$.

This concludes the proof of Assertion 7.2, since the trajectory $y^{(i)}(t; \varepsilon, \lambda)$ corresponds to the piecewise-constant admissible control $u^{(i)}(t; \varepsilon, \lambda)$.

7.4. Proof of the Maximum Principle*

After the cone of variations has been constructed and its basic property as expressed in Assertion 7.2 has been established, the maximum principle can be obtained in a relatively easy manner—as we shall now show—from the theorem on the existence of a support hyperplane passing through an arbitrary boundary point of a convex subset of R^n.

Assume that $\tilde{\mu}_t$ and $\tilde{x}(t)$, $t_1 \leq t \leq t_2$, are optimal in the weakened sense. Then it follows from Assertion 7.2 that the vertex of the cone of variations (7.6) is a boundary point of Q. Therefore, one can draw at least one support hyperplane to Q through the origin. Let a unit vector ξ be orthogonal to a support hyperplane passing through the origin, and let this vector be directed from Q. Then [see (7.6)]

$$\xi \cdot (\delta x(t_2; \delta\mu_\theta) - \vartheta F) \leq 0 \qquad \forall \delta\mu_t \in \delta\mathfrak{M}_{\tilde{\mu}_t} \quad \text{and} \quad \forall \vartheta \geq 0,$$

or, since $\delta\mu_t$ and ϑ are independent and since $\delta\mu_t$ is contained only in the first term and ϑ only in the second,

$$\xi \cdot \delta x(t_2; \delta\mu_\theta) = \xi \cdot \Gamma(t_2) \int_{t_1}^{t_2} G(t)\langle \delta\mu_t, f(t, \tilde{x}(t), u)\rangle \, dt \leq 0 \qquad \forall \delta\mu_t \in \delta\mathfrak{M}_{\tilde{\mu}_t}, \tag{7.13}$$

$$\xi \cdot F \geq 0.$$

The matrix $G(t)$ satisfies the differential equation (see Chapter 5)

$$\dot{G}(t) = -G(t)F_x(t, \tilde{x}(t)).$$

Therefore, the n-dimensional row

$$\tilde{\psi}(t) = \xi \cdot \Gamma(t_2)G(t), \qquad t_1 \leq t \leq t_2,$$

satisfies the equation

$$\dot{\tilde{\psi}}(t) = -\tilde{\psi}(t)F_x(t, \tilde{x}(t)) = -\tilde{\psi}(t)\langle \tilde{\mu}_t, f_x(t, \tilde{x}(t), u)\rangle$$

and the boundary condition

$$\tilde{\psi}(t_2) = \xi \cdot \Gamma(t_2)G(t_2) = \xi \neq 0.$$

With the aid of the function $\tilde{\psi}(t)$, the first of the inequalities (7.13) can

---

*It can be seen from the proof presented that the maximum principle holds not only for solutions $\tilde{\mu}_t, \tilde{x}(t)$ which are optimal in the weakened sense, but which are optimal even under a weaker assumption. Namely, it holds when the generalized control $\tilde{\mu}_t$ is better than all piecewise-constant admissible controls, which constitute a class narrower than Ω_U.

Proof of the Maximum Principle

be rewritten in the form of the following inequality:

$$\int_{t_1}^{t_2} \langle \delta\mu_t, \tilde{\psi}(t) f(t, \tilde{x}(t), u)\rangle \, dt \leq 0 \qquad \forall \delta\mu_t \in \delta\mathfrak{M}_{\tilde{\mu}_t}.$$

This inequality expresses the maximum principle in integral form. Indeed, since

$$\delta\mu_t = \mu_t - \tilde{\mu}_t \in \delta\mathfrak{M}_{\tilde{\mu}_t}$$

is arbitrary, we have

$$\int_{t_1}^{t_2} \langle \tilde{\mu}_t, H(t, \tilde{x}(t), \tilde{\psi}(t), u)\rangle \, dt \geq \int_{t_1}^{t_2} \langle \mu_t, H(t, \tilde{x}(t), \tilde{\psi}(t), u)\rangle \, dt \qquad \forall \mu_t \in \mathfrak{M}_U.$$

Thus, the function $\tilde{\psi}(t) \neq 0$, $t_1 \leq t \leq t_2$, together with the functions $\tilde{\mu}_t$ and $\tilde{x}(t)$ on the same time interval, give an extremal of the convex optimal problem (7.4),

$$\dot{\tilde{x}}(t) = \langle \tilde{\mu}_t, f(t, \tilde{x}(t), u)\rangle = \frac{\partial}{\partial \psi} \langle \tilde{\mu}_t, H(t, \tilde{x}(t), \tilde{\psi}(t), u)\rangle,$$

$$\dot{\tilde{\psi}}(t) = -\tilde{\psi}(t)\langle \tilde{\mu}_t, f_x(t, \tilde{x}(t), u)\rangle = -\frac{\partial}{\partial x} \langle \tilde{\mu}_t, H(t, \tilde{x}(t), \tilde{\psi}(t), u)\rangle, \qquad (7.14)$$

$$\int_{t_1}^{t_2} \langle \tilde{\mu}_t, H(t, \tilde{x}(t), \tilde{\psi}(t), u)\rangle \, dt \geq \int_{t_1}^{t_2} \langle \mu_t, H(t, \tilde{x}(t), \tilde{\psi}(t), u)\rangle \, dt \qquad \forall \mu_t \in \mathfrak{M}_U.$$

It remains to prove that the function

$$\sup_{u \in U} H(t, \tilde{x}(t), \tilde{\psi}(t), u) = M(t, \tilde{x}(t), \tilde{\psi}(t)), \qquad t_1 \leq t \leq t_2,$$

is continuous on the entire interval $t_1 \leq t \leq t_2$ and that

$$M(t_2, \tilde{x}(t_2), \tilde{\psi}(t_2)) \geq 0.$$

The continuity can be obtained as a consequence of the maximum condition. Indeed, in deriving the implication (7.2) $\Rightarrow$ (7.3) we have shown that, if $\tilde{\mu}_t$ satisfies the maximum condition in integral form, then the equality

$$\langle \tilde{\mu}_\theta, H(\theta, \tilde{x}(\theta), \tilde{\psi}(\theta), u)\rangle = M(\theta, \tilde{x}(\theta), \tilde{\psi}(\theta)) \qquad (7.15)$$

holds at every Lebesgue point θ of the function

$$F(t, \tilde{x}(t)) = \langle \tilde{\mu}_t, f(t, \tilde{x}(t), u)\rangle.$$

Since all the measures $\bar{\mu}_t$, $t \in R$, are, by assumption, concentrated on the same bounded set

$$N \subset U \subset R^r,$$

we have

$$\tilde{\psi}(\theta)\langle \bar{\mu}_\theta, f(\theta, \tilde{x}(\theta), u)\rangle = \langle \bar{\mu}_\theta, H(\theta, \tilde{x}(\theta), \tilde{\psi}(\theta), u)\rangle \leq \sup_{u \in \bar{N}} H(\theta, \tilde{x}(\theta), \tilde{\psi}(\theta), u)$$

$$= \max_{u \in \bar{N}} H(\theta, \tilde{x}(\theta), \tilde{\psi}(\theta), u),$$

because the closure $\bar{N}$ is compact. But

$$\max_{u \in \bar{N}} H(\theta, \tilde{x}(\theta), \tilde{\psi}(\theta), u) \leq \sup_{u \in U} H(\theta, \tilde{x}(\theta), \tilde{\psi}(\theta), u) = M(\theta, \tilde{x}(\theta), \tilde{\psi}(\theta)).$$

Therefore, by virtue of (7.15), we have

$$M(\theta, \tilde{x}(\theta), \tilde{\psi}(\theta)) = \max_{u \in \bar{N}} H(\theta, \tilde{x}(\theta), \tilde{\psi}(\theta), u) \tag{7.16}$$

at every Lebesgue point θ of the function $F(t, \tilde{x}(t))$.

The function

$$M_{\bar{N}}(t, x, \psi) = \max_{u \in \bar{N}} H(t, x, \psi, u)$$

depends continuously on (t, x, ψ).* It follows that the equality (7.16), which can be rewritten in the form

$$M(\theta, \tilde{x}(\theta), \tilde{\psi}(\theta)) = M_{\bar{N}}(\theta, \tilde{x}(\theta), \tilde{\psi}(\theta)),$$

holds at all points of the interval $[t_1, t_2]$. Indeed, if for some point $\hat{t}$ we have

$$M_{\bar{N}}(\hat{t}, \tilde{x}(\hat{t}), \tilde{\psi}(\hat{t})) < M(\hat{t}, \tilde{x}(\hat{t}), \tilde{\psi}(\hat{t})) = \sup_{u \in U} H(\hat{t}, \tilde{x}(\hat{t}), \tilde{\psi}(\hat{t}), u),$$

---

*Let us show that the function $M_N(t, x, \psi)$ is upper semicontinuous at an arbitrary point (t, x, ψ). Assume the contrary. Then, by virtue of the continuity of $H(t, x, \psi, u)$ and the compactness of $\bar{N}$, there exist sequences δt_i, δx_i, and $\delta \psi_i$ converging to zero, a convergent sequence of points $u_i \in \bar{N}$, $u_i \to \hat{u}$ $(i \to \infty)$, and a positive value $\varepsilon > 0$ such that

$$M_N(t + \delta t_i, x + \delta x_i, \psi + \delta \psi_i) = H(t + \delta t_i, x + \delta x_i, \psi + \delta \psi_i, u_i) \geq M(t, x, \psi) + \varepsilon \geq H(t, x, \psi, \hat{u}) + \varepsilon.$$

Passing to the limit in these relations as $i \to \infty$, we obtain the contradiction

$$H(t, x, \psi, \hat{u}) \geq H(t, x, \psi, \hat{u}) + \varepsilon.$$

The lower semicontinuity can be proved in a similar way (without the assumption that $\bar{N}$ is compact).

Proof of the Maximum Principle

then there exists a point $\hat{u} \in U$ such that

$$M_{\bar{N}}(\hat{t}, \tilde{x}(\hat{t}), \tilde{\psi}(\hat{t})) < H(\hat{t}, \tilde{x}(\hat{t}), \tilde{\psi}(\hat{t}), \hat{u}).$$

Since both $M_{\bar{N}}(t, \tilde{x}(t), \tilde{\psi}(t))$ and $H(t, \tilde{x}(t), \tilde{\psi}(t), \hat{u})$ ($\hat{u}$ is fixed) depend continuously on t, we obtain an inequality which contradicts the equality (7.16) for all Lebesgue points θ which are close to $\hat{t}$. This completes the proof of continuity.

In deriving the relations (7.14) and proving the continuity of $M(t, \tilde{x}(t), \tilde{\psi}(t))$, we have thus far not made use of the second one of the inequalities (7.13). This is equivalent to using only that part of the cone of variations which is obtained when one sets $\vartheta = 0$ in (7.6).

In order to prove the last relation of the theorem, i.e.,

$$M(t_2, \tilde{x}(t_2), \tilde{\psi}(t_2)) \geq 0,$$

we must make use of the entire cone of variations Q and employ the second of the inequalities (7.13),

$$\xi \cdot F \geq 0.$$

By virtue of the equality $\tilde{\psi}(t_2) = \xi$, this inequality can be rewritten in the form

$$\tilde{\psi}(t_2) \cdot F \geq 0.$$

Let

$$F = \frac{\tilde{x}(t_2) - \tilde{x}(t_2 - \eta_j)}{\eta_j} + v_j, \qquad \eta_j \to 0, \, v_j \to 0 \, (j \to \infty).$$

This equality can be written in the form

$$F = \frac{1}{\eta_j} \int_{t_2 - \eta_j}^{t_2} \dot{\tilde{x}}(t) \, dt + v_j = \frac{1}{\eta_j} \int_{t_2 - \eta_j}^{t_2} \langle \bar{\mu}_t, f(t, \tilde{x}(t), u) \rangle \, dt + v_j.$$

Hence

$$\tilde{\psi}(t_2) \cdot F = \frac{1}{\eta_j} \int_{t_2 - \eta_j}^{t_2} \tilde{\psi}(t) \langle \bar{\mu}_t, f(t, \tilde{x}(t), u) \rangle \, dt$$

$$+ \frac{1}{\eta_j} \int_{t_2 - \eta_j}^{t_2} (\tilde{\psi}(t_2) - \tilde{\psi}(t)) \langle \bar{\mu}_t, f(t, \tilde{x}(t), u) \rangle \, dt + v_j.$$

Making use of the equivalence of the maximum conditions (7.2) and

(7.3), and substituting the function $M(t, \tilde{x}(t), \tilde{\psi}(t))$, which is equal to it almost everywhere, for the integrand $\tilde{\psi}(t)\langle\tilde{\mu}_t, f(t, \tilde{x}(t), u)\rangle$, we obtain

$$\left|\tilde{\psi}(t_2)\cdot F - \frac{1}{\eta_j}\int_{t_2-\eta_j}^{t_2} M(t, \tilde{x}(t), \tilde{\psi}(t))\, dt\right| \leq \max_{t_2-\eta_j \leq t \leq t_2} |\tilde{\psi}(t_2)-\tilde{\psi}(t)|$$

$$\times \frac{1}{\eta_j}\int_{t_2-\eta_j}^{t_2} |\langle\tilde{\mu}_t, f(t, \tilde{x}(t), u)\rangle|\, dt + v_j.$$

Since the measures $\tilde{\mu}_t$ are concentrated on the same bounded set, and since the function $f(t, x, u)$ is continuous, we have

$$\max_{t_2-\eta_j \leq t \leq t_2} |\langle\tilde{\mu}_t, f(t, \tilde{x}(t), u)\rangle| \leq \text{const},$$

and we obtain the estimate

$$\left|\tilde{\psi}(t_2)\cdot F - \frac{1}{\eta_j}\int_{t_2-\eta_j}^{t_2} M(t, \tilde{x}(t), \tilde{\psi}(t))\, dt\right| \leq \max_{t_2-\eta_j \leq t \leq t_2} |\tilde{\psi}(t_2)-\tilde{\psi}(t)|\text{const} + v_j.$$

Passing to the limit as $j \to \infty$, we obtain from the continuity of $M(t, \tilde{x}(t), \tilde{\psi}(t))$ that

$$M(t_2, \tilde{x}(t_2), \tilde{\psi}(t_2)) = \tilde{\psi}(t_2)\cdot F \geq 0.$$

In conclusion, we note that the following statement is seen to hold from the proof: The function $\tilde{\psi}(t)$, $t_1 \leq t \leq t_2$, can be taken to be an arbitrary non-zero solution of the differential equation

$$\dot{\psi} = -\psi F_x(t, \tilde{x}(t))$$

that satisfies the following boundary condition: The vector $\tilde{\psi}(t_2)$ is orthogonal to a support hyperplane to Q which passes through the origin and is directed from Q.

Note that the boundary condition for $\tilde{\psi}(t)$ is given at the final time t_2, and not at the initial time $t = t_1$.

8

The Existence of Optimal Solutions

The advantages of generalized controls over ordinary controls and of the corresponding passage to the convex control problem were very apparent in Chapter 6, where we varied the trajectories of controlled equations, and in Chapter 7, where we proved the maximum principle.

Another important advantage of generalized controls becomes very clear while studying the question of the existence of optimal solutions. This is related to the fact that, under very broad and natural assumptions, the set of generalized controls turns out to be compact in the topology of weak convergence (the precise definitions are given below). Compactness, in turn, implies the existence of solutions for a large class of convex optimal problems. On the other hand, the ordinary optimal problem (1.3) does not have a solution unless special restrictions are imposed on the right-hand side $f(t, x, u)$ (see Section 8.3).

We shall begin by proving a general assertion about the weak compactness of a set in the space of Radon measures (Assertion 8.1) and a corollary to this assertion, namely, the theorem on the weak compactness of the class of generalized controls $\mathfrak{M}_U$ for compact sets $U \subset R^r$ (Theorem 8.1).

Theorem 8.2 on the existence of solutions in a very general convex optimal problem follows easily from Theorem 8.1. Using this theorem, we shall prove Theorem 8.3 on the existence of optimal solutions in the class of ordinary controls (the A. F. Filippov theorem).

We shall then consider applications of Theorem 8.3 to sliding optimal regimes and, as another application, we shall prove the Hilbert–Tonelli existence theorem (Theorem 8.4) of the classical calculus of variations.

8.1. The Weak Compactness of the Class of Generalized Controls

In this chapter, $C^0(R^m)$ denotes the linear space of all continuous functions on R^m with compact supports, $\|\cdot\|_M$ denotes the seminorm on $C^0(R^m)$ determined by a subset $M \subset R^m$ according to the formula

$$\|g(z)\|_M = \sup_{z \in M} |g(z)|, \qquad g(z) \in C^0(R^m),$$

and $C^*(R^m)$ denotes the set of all real (not necessarily positive) Radon measures on R^m.

In accordance with the definition given in Chapter 2, a sequence of measures $v^{(i)} \in C^*(R^m)$, $i = 1, 2, \ldots$, is said to converge weakly to a measure $v \in C^*(R^m)$ if

$$\langle v^{(i)}, g(z) \rangle = \int_{R^m} g(z) \, dv^{(i)}(z) \to \langle v, g(z) \rangle = \int_{R^m} g(z) \, dv(z) \qquad (i \to \infty)$$

for any function $g(z) \in C^0(R^m)$.

We denote by $C^*(M)$ the subspace of the space $C^*(R^m)$ which consists of all Radon measures concentrated on the set $M \subset R^m$.

It is easy to see that, if the set M is closed, then the subspace $C^*(M)$ is closed (sequentially) with respect to weak convergence, i.e., the relation $v^{(i)} \in C^*(M)$, $i = 1, 2, \ldots$, and the weak convergence $v^{(i)} \to v$ $(i \to \infty)$ imply $v \in C^*(M)$.

Indeed, since the complement $R^m \setminus M$ is open, it is enough to show that the equality $\langle v, g(z) \rangle = 0$ holds for any function $g(z) \in C^0(R^m)$ whose support is contained in $R^m \setminus M$. However, this is obvious, since we have $\langle v^{(i)}, g(z) \rangle = 0$ for all $i = 1, 2, \ldots$, so that

$$\langle v, g(z) \rangle = \langle v, g(z) \rangle - \lim_{i \to \infty} \langle v^{(i)}, g(z) \rangle = \lim_{i \to \infty} \langle v - v^{(i)}, g(z) \rangle = 0.$$

The assertion becomes invalid if the set M is not closed. One can easily convince oneself that this is indeed so with the following simple example: We take a sequence of points z_i which converges to a point z distinct from all z_i. Then the sequence of Dirac measures δ_{z_i} converges weakly to the measure δ_z and, taking for M the union of all z_i, we see that the measure δ_z is concentrated at the point $z \notin M$.

As before, we use the notation $\|v\|$ to represent the norm (full variation) of a measure v. The following obvious inequality holds:

$$|\langle v, g(z) \rangle| \leq \|v\| \cdot \|g(z)\|_M \qquad \forall v \in C^*(M).$$

The Existence of Optimal Solutions

We shall say that a set $\mathfrak{N} \subset C^*(R^m)$ is *weakly (sequentially) compact* if, from an arbitrary sequence of measures $v^{(i)} \in \mathfrak{N}$, $i = 1, 2, \ldots$, one can choose a subsequence which converges weakly to a measure that also belongs to $\mathfrak{N}$.

Assertion 8.1. For an arbitrary closed set $M \subset R^m$ and an arbitrary nonnegative constant, which we shall denote as const, the corresponding set of Radon measures

$$\mathfrak{N} = \mathfrak{N}(M, \text{const}) = \{v : v \in C^*(M), \|v\| \leq \text{const}\} \subset C^*(R^m)$$

is weakly compact.

Proof. In the linear space $C^0(R^m)$, we consider the seminorm $\|\cdot\|_M$. We choose a sequence of functions $g_1(z), g_2(z), \ldots$ which is everywhere dense in $C^0(R^m)$ with respect to this seminorm, i.e., a sequence such that, for every $g(z) \in C^0(R^m)$ and every $\varepsilon > 0$, there exists a $g_i(z)$ such that

$$\|g(z) - g_i(z)\|_M \leq \varepsilon.$$

The existence of a sequence $g_1(z), g_2(z), \ldots$, is a direct consequence of Assertion 3.3 and the inequality

$$\|g(z)\|_M \leq \|g(z)\| = \sup_{z \in R^m} |g(z)|,$$

which holds for an arbitrary set M.

Since the sequence of values

$$|\langle v^{(i)}, g_1(z)\rangle| \leq \|v^{(i)}\| \cdot \|g_1(z)\|_M \leq \text{const} \, \|g_1(z)\|_M, \qquad i = 1, 2, \ldots,$$

is bounded, we can choose a subsequence $v^{(i_k)} = v_1^{(k)}$, $k = 1, 2, \ldots$, such that there exists a limit

$$\lim_{i \to \infty} \langle v_1^{(i)}, g_1(z)\rangle = \lambda_1.$$

By similar considerations, one can choose from the sequence $v_1^{(i)}$ a subsequence

$$v_2^{(k)} = v_1^{(i_k)}, \qquad k = 1, 2, \ldots,$$

such that there exists a limit

$$\lim_{i \to \infty} \langle v_2^{(i)}, g_2(z)\rangle = \lambda_2,$$

etc. We obtain the system of embedded sequences

$$\{v_1^{(i)}\}_1^\infty \supset \{v_2^{(i)}\}_1^\infty \supset \cdots,$$

that satisfy the condition

$$\lim_{i\to\infty} \langle v_k^{(i)}, g_j(z)\rangle = \lambda_j \qquad \forall k \geq j, j=1, 2, \ldots.$$

Choosing the diagonal sequence

$$\hat{v}^{(i)} = v_i^{(i)}, \qquad i=1, 2, \ldots,$$

we obtain

$$\lim_{i\to\infty} \langle \hat{v}^{(i)}, g_j(z)\rangle = \lambda_j \qquad \forall j=1, 2, \ldots.$$

We shall show that the sequence of values $\langle \hat{v}^{(i)}, g(z)\rangle$ converges as $i\to\infty$ for every function $g(z) \in C^0(R^m)$. Since the sequence of functions $g_j(z)$, $j=1, 2, \ldots$, is everywhere dense in $C^0(R^m)$ with respect to the seminorm $\|\cdot\|_M$, for every $\varepsilon > 0$, there exists a $j = j_\varepsilon$ such that

$$\|g(z) - g_{j_\varepsilon}(z)\|_M \leq \frac{\varepsilon}{4\cdot\text{const}}.$$

Also, since the sequence of values $\langle \hat{v}^{(i)}, g_{j_\varepsilon}(z)\rangle$ converges as $i\to\infty$, we have for i', i'' sufficiently large

$$|\langle \hat{v}^{(i')} - \hat{v}^{(i'')}, g_{j_\varepsilon}(z)\rangle| \leq \frac{\varepsilon}{2}.$$

Since

$$\|\hat{v}^{(i')} - \hat{v}^{(i'')}\| \leq \|\hat{v}^{(i')}\| + \|\hat{v}^{(i'')}\| \leq 2\cdot\text{const},$$

it follows that, for every $\varepsilon > 0$, there exist a j_ε and an i_1 such that

$$|\langle \hat{v}^{(i')}, g(z)\rangle - \langle \hat{v}^{(i'')}, g(z)\rangle| \leq |\langle \hat{v}^{(i')} - \hat{v}^{(i'')}, g(z) - g_{j_\varepsilon}(z)\rangle| + |\langle \hat{v}^{(i')} - \hat{v}^{(i'')}, g_{j_\varepsilon}(z)\rangle|$$

$$\leq \|\hat{v}^{(i')} - \hat{v}^{(i'')}\| \frac{\varepsilon}{4\cdot\text{const}} + \frac{\varepsilon}{2} \leq \varepsilon$$

for $i', i'' \geq i_1$. Thus, the sequence of values $\langle \hat{v}^{(i)}, g(z)\rangle$, $i=1, 2, \ldots$, converges as $i\to\infty$ for every $g(z) \in C^0(R^m)$, and we can define the functional $v(g(z))$ on $C^0(R^m)$ by the formula

$$g(z) \to v(g(z)) = \lim_{i\to\infty} \langle \hat{v}^{(i)}, g(z)\rangle.$$

Obviously, for every $g'(z), g''(z) \in C^0(R^m)$ and any real numbers α and β, we have

$$v(\alpha g'(z) + \beta g''(z)) = \lim_{i\to\infty} \langle \hat{v}^{(i)}, \alpha g'(z) + \beta g''(z)\rangle$$

$$= \alpha \lim_{i\to\infty} \langle \hat{v}^{(i)}, g'(z)\rangle + \beta \lim_{i\to\infty} \langle \hat{v}^{(i)}, g''(z)\rangle = \alpha v(g'(z)) + \beta v(g''(z))$$

The Existence of Optimal Solutions

and

$$|v(g'(z))| = \lim_{i \to \infty} |\langle \hat{v}^{(i)}, g'(z) \rangle| \leq \lim_{i \to \infty} \sup \|\hat{v}^{(i)}\| \cdot \|g'(z)\|_M \leq \text{const } \|g'(z)\|_M.$$

It follows that v is a linear functional on the linear space $C^0(R^m)$, and that its restriction to an arbitrary subspace of functions from $C^0(R^m)$ whose supports are contained in a given compact set $K \subset R^m$ is continuous in the norm of uniform convergence on K. In other words, v is a Radon measure on R^m.

We write

$$v(g(z)) = \langle v, g(z) \rangle.$$

According to the definition of v, we have

$$\lim_{i \to \infty} \langle v - \hat{v}^{(i)}, g(z) \rangle = 0 \qquad \forall g(z) \in C^0(R^m),$$

i.e., the subsequence $\hat{v}^{(i)}$ of the sequence $v_i \in \mathfrak{R}(M, \text{const})$ converges weakly to the measure v as $i \to \infty$.

The estimate

$$|v(g(z))| = |\langle v, g(z) \rangle| \leq \text{const } \|g(z)\|_M \qquad \forall g(z) \in C^0(R^m)$$

yields

$$\|v\| \leq \text{const}.$$

Since the set $C^*(M)$ is weakly closed (because M is closed) and all $\hat{v}^{(i)} \in C^*(M)$, we obtain the relation

$$v \in \mathfrak{R}(M, \text{const}),$$

which completes the proof of the assertion.

Remark. If the closed set M is not compact, then all measures of the sequence $\hat{v}^{(i)}$, $i = 1, 2, \ldots$, can be probability measures and, nevertheless, the limit measure v can be zero.

For example, if $M = R^m$ and the sequence of points $z_i \in R^m$ goes to infinity, then the sequence of the Dirac measures δ_{z_i}, $i = 1, 2, \ldots$, will converge to the zero measure. This is so because

$$\langle \delta_{z_i}, g(z) \rangle = 0$$

for those i for which z_i lies outside of the support of the function $g(z)$.

On the other hand, if M is compact, then such a "loss of measure at infinity" cannot take place. It is also not difficult to show that if all measures of the sequence are probability measures, then the limit measure will also be a probability measure (the measure is preserved in the limit).

We now pass to generalized controls $\mu_t \in \mathfrak{M}_U$. As we said in Chapter 2, we consider these controls as Radon measures v on R^{1+r} which are defined by the formula

$$\langle v, g(t, u) \rangle = \int_R \langle \mu_t, g(t, u) \rangle \, dt = \int_R dt \int_{R^r} g(t, u) \, d\mu_t(u).$$

We shall prove the following basic theorem:

Theorem 8.1. If the set of admissible values $U \subset R^r$ is compact, then the set of all generalized controls $\mathfrak{M}_U$ is weakly (sequentially) compact, i.e., from an arbitrary sequence of generalized controls $\mu_t^{(i)} \in \mathfrak{M}_U$, $i = 1, 2, \ldots$, one can choose a subsequence which converges weakly to a generalized control $\mu_t \in \mathfrak{M}_U$.*

Proof. Given a sequence of generalized controls $\mu_t^{(i)}$, $i = 1, 2, \ldots$, we define the sequence of Radon measures on R^{1+r} which depend on a parameter $\tau \in R$,

$$v_\tau^{(i)}, \qquad i = 1, 2, \ldots,$$

by the formula

$$[v_\tau^{(i)}, g(t, u)] = \int_{R^{1+r}} g(t, u) \, dv_\tau^{(i)}(t, u) = \int_0^\tau \langle \mu_t^{(i)}, g(t, u) \rangle \, dt$$

$$= \int_0^\tau dt \int_U g(t, u) \, d\mu_t^{(i)}(u), \qquad g(t, u) \in C^0(R^{1+r}). \tag{8.1}$$

The "scalar product" (the result of the action) of the measure $v_\tau^{(i)}$ with (on) the function $g(t, u)$ will be denoted, as in (8.1), by square brackets $[\cdot]$. The angle brackets $\langle \cdot \rangle$ will retain their original meaning, i.e., the integration with respect to measure μ_t (with fixed t) of a function of points $u \in R^r$.

It is easy to see that, for any i, the measure $v_\tau^{(i)}$ is concentrated on the compact set $[0, \tau] \times U \subset R^{1+r}$. Indeed, if the support of the function

---

*Thus, in the case of a compact $U \subset R^r$, an arbitrary set of measurable functions $u(t)$, $t_1 \leq t \leq t_2$, which takes on values in U is relatively compact in the topology of weak convergence of generalized controls.

The Existence of Optimal Solutions

$g(t, u)$ lies in $R^{1+r} \backslash [0, \tau] \times U$, then

$$g(t, u) = 0 \quad \forall (t, u) \in [0, \tau] \times U$$

and, therefore,

$$[v_\tau^{(i)}, g(t, u)] = \int_0^\tau dt \int_U g(t, u) \, d\mu_t^{(i)} = 0.$$

Further, it is clear that

$$\|v_\tau^{(i)}\| = \left| \int_0^\tau \|\mu_t^{(i)}\| \, dt \right| = |\tau|.$$

Taking into account the form of the formula (8.1), we shall write the measure $v_\tau^{(i)}$ in the form

$$dv_\tau^{(i)} = d\mu_t^{(i)} \, dt, \quad t \in [0, \tau], \tau \in R.$$

Let $\tau_1, \tau_2, \ldots$ be a sequence of points which is everywhere dense in R. With the aid of this sequence, we construct a double sequence of Radon measures on R^{1+r}:

$$dv_{\tau_j}^{(i)} = d\mu_t^{(i)} \, dt, \quad t \in [0, \tau_j], i, j = 1, 2, \ldots.$$

By virtue of Assertion 8.1, for every τ_j one can choose from this sequence a subsequence that converges weakly to a measure that is concentrated on $[0, \tau_j] \times U$ and whose norm does not exceed $|\tau_j|$. Therefore, the diagonal process that we employed in the proof of Assertion 8.1 will allow us to choose a sequence of integers $i_1 < i_2 < \cdots$ such that, for every fixed j and for $k \to \infty$, the sequence of measures

$$v_{\tau_j}^{(i_k)}, \quad k = 1, 2, \ldots,$$

converges weakly to a Radon measure v_{τ_j} concentrated on $[0, \tau_j] \times U$.

It is easy to see that the sequence of measures $v_\tau^{(i_k)}, k = 1, 2, \ldots$, converges weakly not only for $\tau = \tau_j, j = 1, 2, \ldots$, but for an arbitrary $\tau \in R$. Indeed, for every $\tau', \tau'' \in R$, and every $g(t, u) \in C^0(R^{1+r})$, we have

$$\left| [v_{\tau'}^{(i_k)}, g(t, u)] - [v_{\tau''}^{(i_k)}, g(t, u)] \right| = \left| \int_{\tau''}^{\tau'} \langle \mu_t^{(i_k)}, g(t, u) \rangle \, dt \right| \leq |\tau' - \tau''| \cdot \|g(t, u)\|_{[\tau', \tau''] \times U}.$$

(8.2)

For every $\tau \in R$ and every $\varepsilon > 0$, there exists a τ_j such that
$$|\tau - \tau_j| \leq \varepsilon.$$
Therefore, we obtain
$$|[v_\tau^{(ik')} - v_\tau^{(ik'')}, g(t,u)]| \leq |[v_\tau^{(ik')} - v_{\tau_j}^{(ik')}, g(t,u)]| + |[v_{\tau_j}^{(ik')} - v_{\tau_j}^{(ik'')}, g(t,u)]|$$
$$+ |[v_\tau^{(ik'')} - v_{\tau_j}^{(ik'')}, g(t,u)]| \leq 2\varepsilon \|g(t,u)\|_{[\tau-\varepsilon, \tau+\varepsilon] \times U} + \varepsilon$$

for all k' and k'' sufficiently large. It follows that the sequence of values $\langle v_\tau^{(ik)}, g(t,u) \rangle$, $k = 1, 2, \ldots$, is fundamental and therefore converges to a limit $v_\tau(g)$.

Repeating the corresponding arguments of the proof of Assertion 8.1, we obtain easily that, for every fixed $\tau \in R$, the mapping
$$g(t,u) \to v_\tau(g)$$
is the Radon measure on R^{1+r} concentrated on $[0, \tau] \times U$ to which the sequence of measures $v_\tau^{(ik)}$ converges weakly as $k \to \infty$.

Thus, we have obtained a family of Radon measures v_τ, $\tau \in R$, on R^{1+r}. Our goal is to prove the existence of a generalized control $\mu_t \in \mathfrak{M}_U$ such that

$$[v_\tau, g(t,u)] = \int_0^\tau \langle \mu_t, g(t,u) \rangle \, dt \quad \forall \tau \in R \text{ and } \forall g(t,u) \in C^0(R^{1+r}), \quad (8.3)$$

or, according to our notation, $dv_\tau = d\mu_t dt$. This will prove the theorem, since—as it is easy to see—in this case the sequence of generalized controls $\mu_t^{(ik)}$ converges weakly to μ_t as $k \to \infty$.

Indeed, if a $\tau > 0$ is so large that the projection of the support of $g(t,u)$ onto the t-axis is contained in $[-\tau, \tau]$, then

$$\int_R \langle \mu_t^{(ik)} - \mu_t, g(t,u) \rangle \, dt = \int_{-\tau}^\tau \langle \mu_t^{(ik)} - \mu_t, g(t,u) \rangle \, dt$$
$$= [v_\tau^{(ik)} - v_\tau, g(t,u)] - [v_{-\tau}^{(ik)} - v_{-\tau}, g(t,u)] \to 0, \quad k \to \infty.$$

Let us prove the representation (8.3). If we pass to the limit as $k \to \infty$ in the estimate (8.2), then we obtain the estimate

$$|[v_{\tau'} - v_{\tau''}, g(t,u)]| \leq |\tau' - \tau''| \cdot \|g(t,u)\|_{[\tau', \tau''] \times U}.$$

This implies that the following function of τ [which depends on the choice of $g(t,u)$],
$$h_g(\tau) = [v_\tau, g(t,u)], \quad \tau \in R,$$

The Existence of Optimal Solutions

is Lipschitzian and therefore absolutely continuous.

Thus, there exists a set of full measure $T_g \subset R$ such that the function $h_g(\tau)$ is differentiable at every point of the set. We denote by $\mu_{\hat{\tau}}(g)$ the derivative of $h_g(\tau)$ at a point of differentiability $\hat{\tau}$,

$$\mu_{\hat{\tau}}(g) = \frac{d}{d\tau}[v_\tau, g(t, u)]\big|_{\tau = \hat{\tau}}.$$

Of course, the set T_g depends on $g(t, u)$, which is noted by the subscript. We shall prove the existence of a set of full measure $T \subset R$ such that *any* function

$$h_g(\tau) = [v_\tau, g(t, u)], \quad \tau \in R, \quad g(t, u) \in C^0(R^{1+r}),$$

is differentiable at all points of the set T. To this end, we choose a sequence of functions

$$g_i(t, u) \in C^0(R^{1+r}), \quad i = 1, 2, \ldots,$$

which is everywhere dense in $C^0(R^{1+r})$ with respect to the seminorm $\|\cdot\|_{R \times U}$, i.e., a sequence that satisfies the following condition: For every $g(t,u) \in C^0(R^{1+r})$ and for every $\varepsilon > 0$, there exists a value i such that

$$\|g(t, u) - g_i(t, u)\|_{R \times U} \leq \varepsilon.$$

The function $g(t, u) \in C^0(R^{1+r})$ was arbitrary when we constructed the function $h_g(\tau)$ with $\tau \in R$. Therefore, we have the sequence of functions

$$h_{g_i}(\tau), \quad \tau \in R, \quad i = 1, 2, \ldots,$$

and the sequence of sets of full measure $T_{g_i} \subset R$, $i = 1, 2, \ldots$, on which the corresponding functions are differentiable. The intersection

$$T = \bigcap_{i=1}^{\infty} T_{g_i} \subset R$$

is a set of full measure. We shall now show that an arbitrary function

$$h_g(\tau), \quad \tau \in R, \quad g(t, u) \in C^0(R^{1+r})$$

is differentiable at all points of the set T. Obviously, in order to prove this it is sufficient to show that, for all $\hat{\tau} \in T$ and for $\tau', \tau'' \to \hat{\tau}$, we have

$$\frac{h_g(\tau') - h_g(\hat{\tau})}{\tau' - \hat{\tau}} - \frac{h_g(\tau'') - h_g(\hat{\tau})}{\tau'' - \hat{\tau}} \to 0. \tag{8.4}$$

Using the notation

$$\Delta_{\tau_1,\tau_2}(g) = \frac{h_g(\tau_1) - h_g(\tau_2)}{\tau_1 - \tau_2} = \frac{[v_{\tau_1} - v_{\tau_2}, g(t, u)]}{\tau_1 - \tau_2},$$

we can write on the basis of (8.2)

$$|\Delta_{\tau',\hat{\tau}}(g) - \Delta_{\tau'',\hat{\tau}}(g)| \leq |\Delta_{\tau',\hat{\tau}}(g - g_i)| + |\Delta_{\tau',\hat{\tau}}(g_i) - \Delta_{\tau'',\hat{\tau}}(g_i)| + |\Delta_{\tau'',\hat{\tau}}(g - g_i)|$$
$$\leq 2\|g(t, u) - g_i(t, u)\|_{R \times U} + |\Delta_{\tau',\hat{\tau}}(g_i) - \Delta_{\tau'',\hat{\tau}}(g_i)|$$

and, since the function $h_{g_i}(\tau)$ is differentiable at $\hat{\tau} \in T$,

$$\lim_{\tau',\tau'' \to \hat{\tau}} \sup |\Delta_{\tau',\hat{\tau}}(g) - \Delta_{\tau'',\hat{\tau}}(g)| \leq 2\|g(t, u) - g_i(t, u)\|_{R \times U}.$$

The left-hand side of this inequality does not depend on $g_i(t, u)$ and the right-hand side can be made arbitrarily small by a proper choice of $g_i(t, u)$, which implies the relation (8.4). Thus, for all $g(t, u) \in C^0(R^{1+r})$, the derivative

$$\mu_\tau(g) = \frac{d[v_\tau, g(t, u)]}{d\tau}$$

exists at all points of the set of full measure $T \subset R$ and determines the family of functionals

$$\mu_\tau: C^0(R^{1+r}) \to R, \qquad \tau \in T,$$

given by the formula

$$g(t, u) \to \mu_\tau(g).$$

Obviously, the functional μ_τ is linear,

$$\mu_\tau(\alpha g'(t, u) + \beta g''(t, u)) = \alpha\mu_\tau(g'(t, u)) + \beta\mu_\tau(g''(t, u)).$$

It is easy to see that

$$|\mu_\tau(g(t, u))| \leq \max_{u \in U} |g(\tau, u)| \qquad \forall \tau \in T, \tag{8.5}$$

because the estimate (8.2) yields

$$|\mu_\tau(g)| = \left|\lim_{\tau' \to \tau} \frac{[v_{\tau'} - v_\tau, g(t, u)]}{\tau' - \tau}\right| \leq \lim_{\tau' \to \infty} \sup \|g(t, u)\|_{[\tau',\tau] \times U} = \max_{u \in U} |g(\tau, u)|.$$

Therefore, μ_τ is a Radon measure on R^{1+r} for all $\tau \in T$. Moreover, the last estimate shows that this measure is concentrated on the set $\{\tau\} \times U \subset R^{1+r}$. For this reason, μ_τ can be considered—and we shall do so—as the Radon measure on R^r which is concentrated on the compact set $U \subset R^r$ and acts on

The Existence of Optimal Solutions

an arbitrary function $g(u) \in C^0(R^r)$ according to the formula

$$\langle \mu_\tau, g(u) \rangle = \frac{d}{d\tau} [v_\tau, h(t)g(u)],$$

where $h(t)$ is a continuous function which has a compact support in R and is equal to 1 at $t = \tau$.

The set $T \subset R$ is of full measure, and the function

$$\langle \mu_t, g(t, u) \rangle = \mu_t(g), \quad t \in T,$$

is measurable for any continuous function $g(t, u) \in C^0(R^{1+r})$. Therefore, appropriately defining μ_t on $R \setminus T$, we obtain a measurable family of Radon measures on R^r which is determined for all $t \in R$. We also have the representation

$$[v_\tau, g(t, u)] = \int_0^\tau \mu_t(g) \, dt = \int_0^\tau \langle \mu_t, g(t, u) \rangle \, dt.$$

In order to finish the proof of the theorem, it remains to show that μ_t is a probability measure for all $t \in T$. If a function $g(t, u) \in C^0(R^{1+r})$ is nonnegative, then $\langle \mu_t^{(i)}, g(t, u) \rangle \geq 0$ for any integer i and $t \in R$. Therefore,

$$[v_\tau^{(i_k)}, g(t, u)] = \int_0^\tau \langle \mu_t^{(i_k)}, g(t, u) \rangle \, dt, \quad k = 1, 2, \ldots,$$

are nondecreasing functions of the variable $\tau \in R$, so that their limit as $k \to \infty$ is also a nondecreasing function $[v_\tau, g(t, u)]$ of $\tau \in R$. Finally, it follows that

$$\frac{d}{d\tau} [v_\tau, g(t, u)]\big|_{\tau = \hat{t}} = \langle \mu_{\hat{t}}, g(\hat{t}, u) \rangle = \mu_{\hat{t}}(g) \geq 0$$

for all $\hat{t} \in T$, i.e., $\mu_{\hat{t}}$ is a nonnegative measure.

So far, we have not made use of the fact that U is a compact subset of R^{1+r}. Had this set been an arbitrary closed set (in order to make use of Assertion 8.1), all we have said would remain in force. Therefore, it may well have turned out that $\mu_t = 0$ for all $t \in R$ (the loss of measure at infinity that we spoke about above). Now, we shall take into account the fact that U is compact and prove that $\|\mu_t\| = 1$ for all $t \in T$.

We take an arbitrary point $\hat{t} \in T$, and let $g(t, u) \in C^0(R^{1+r})$ be a nonnegative function contained between 0 and 1 and identically 1 on a neighborhood of the set

$$\{\hat{t}\} \times U \subset R^{1+r}.$$

Such a function exists because U is compact. Then we have $\langle \mu_{\hat{t}}^{(i)}, g(t, u)\rangle = 1$ for all i and all t which are sufficiently close to $\hat{t}$. Hence

$$[v_\tau^{(i)}, g(t, u)] = \int_0^\tau \langle \mu_t^{(i)}, g(t, u)\rangle \, dt = \int_0^{\hat{t}} \langle \mu_t^{(i)}, g(t, u)\rangle \, dt + \int_{\hat{t}}^\tau \langle \mu_t^{(i)}, g(t, u)\rangle \, dt$$

$$= [v_{\hat{t}}^{(i)}, g(t, u)] + (\tau - \hat{t})$$

for all τ sufficiently close to $\hat{t}$. Therefore, for τ under consideration, we have

$$[v_\tau, g(t, u)] = \text{const} + (\tau - \hat{t})$$

and

$$\frac{d}{d\tau}[v_\tau, g(t, u)]\bigg|_{\tau=\hat{t}} = \frac{d\tau}{d\tau} = \langle \mu_{\hat{t}}, g(t, u)\rangle = 1.$$

Thus, $\|\mu_i\| \geqslant 1$, and the inequality (8.5) implies the opposite estimate $\|\mu_i\| \leqslant 1$, so that $\|\mu_i\| = 1$. This means that μ_i is a probability measure.

8.2. The Existence Theorem for Convex Optimal Problems

We shall say that the controlled equation

$$\dot{x} = f(t, x, u), \qquad u \in U,$$

has the property of *unlimited extendability of solutions* if, for an arbitrary generalized control $\mu_t \in \mathfrak{M}_U$ and arbitrary initial data τ and x_τ, the differential equation

$$\dot{x} = \langle \mu_t, f(t, x, u)\rangle$$

has a solution $x(t)$ that satisfies the initial condition

$$x(\tau) = x_\tau$$

and is defined on an arbitrary interval of the time axis.

With the aid of the technique developed in Chapter 4, it is not difficult to show that a controlled equation will have this property if the following

The Existence of Optimal Solutions

estimate holds for $f(t, x, u)$:

$$|f(t, x, u)| \leq m(t)g(|x|) \qquad \forall (t, x, u) \in R \times R^n \times U,$$

where $m(t)$ with $t \in R$ is a locally integrable function, and the continuous function g increases to infinity no faster than $|x|$, i.e.,

$$\limsup_{x \to \infty} \frac{g(|x|)}{|x|} < \infty.$$

Theorem 8.2. Consider the following convex optimal problem,

$$\dot{x} = \langle \mu_t, f(t, x, u) \rangle, \qquad \mu_t \in \mathfrak{M}_U,$$
$$t = t_1, \qquad x(t_1) = x_1, \qquad x(t_2) = x_2, \qquad t_2 - t_1 \to \min, \tag{8.6}$$

which has the property of unlimited extendability of solutions. Assume that the set of admissible values $U \subset R^r$ is compact and that there exists at least one (not necessarily optimal) solution

$$\hat{\mu}_t, \hat{x}(t), \qquad t_1 \leq t \leq \hat{t}_2,$$

of the convex control problem

$$\dot{x} = \langle \mu_t, f(t, x, u) \rangle, \qquad \mu_t \in \mathfrak{M}_U, t = t_1, \qquad x(t_1) = x_1, \qquad x(t_2) = x_2.$$

Then there also exists an optimal solution

$$\tilde{\mu}_t, \tilde{x}(t), \qquad t_1 \leq t \leq t_2,$$

of this control problem, i.e., there exists a solution of the optimal problem (8.6).

Proof. We denote by $\mathfrak{M}_{x_2} \subset \mathfrak{M}_U$ the set of all generalized controls that satisfy the following condition: If $\mu_t \in \mathfrak{M}_{x_2}$, then the corresponding solution

$$x(t; \mu_\theta), \qquad t_1 \leq t \leq \hat{t}_2, \qquad x(t_1; \mu_\theta) = x_1$$

(which is determined, by definition, on the entire interval $[t_1, \hat{t}_2]$) takes on the value x_2 for at least one $t \in [t_1, \hat{t}_2]$. In other words, the curve

$$\{(t, x(t; \mu_\theta)): t \in [t_1, \hat{t}_2]\} \subset R^{1+n}$$

has a nonempty intersection with the interval

$$\{(t, x_2): t \in [t_1, \hat{t}_2]\} \subset R^{1+n}.$$

The set $\mathfrak{M}_{x_2}$ is not empty, because $\hat{\mu}_t \in \mathfrak{M}_{x_2}$ by assumption. For all

$\mu_t \in \mathfrak{M}_{x_2}$, we denote by $\tau(\mu_t)$ the lower limit of those $t \in [t_1, \hat{t}_2]$ for which $x(t; \mu_\theta) = x_2$. By virtue of continuity in t, the solution $x(t; \mu_\theta)$ takes on the value x_2 at $t = \tau(\mu_\theta)$.

Obviously, the theorem will be proved if we show that the numerical-valued function

$$\tau(\mu_t), \quad \mu_t \in \mathfrak{M}_{x_2}, \tag{8.7}$$

attains its minimal value on $\mathfrak{M}_{x_2}$. Then the generalized control $\tilde{\mu}_t$ at which the minimal value $\tilde{\tau} = \tau(\tilde{\mu}_t)$ is attained, and the corresponding trajectory

$$\tilde{x}(t) = x(t; \tilde{\mu}_\theta), \quad t_1 \leq t \leq \tilde{\tau},$$

yields the solution of the optimal problem (8.6); and $\tilde{\tau} - t_1$ will be the optimal transfer time.

In order to prove the existence of such a control $\tilde{\mu}_t$, we shall prove that the set $\mathfrak{M}_{x_2}$ is closed and, therefore, compact (by virtue of the compactness of $\mathfrak{M}_U$, see Theorem 8.1) with respect to weak convergence of generalized controls. We shall also prove that the function (8.7) is lower semicontinuous with respect to this convergence. This means that, if a sequence of generalized controls $\mu_t^{(i)} \in \mathfrak{M}_{x_2}$ converges weakly to a generalized control $\mu_t \in \mathfrak{M}_{x_2}$ as $i \to \infty$, then

$$\liminf_{i \to \infty} \tau(\mu_t^{(i)}) \geq \tau(\mu_t).$$

Let

$$\mu_t^{(i)} \in \mathfrak{M}_{x_2}, \quad \mu_t^{(i)} \to \mu_t \quad (i \to \infty).$$

A solution

$$x(t) = x(t; \mu_\theta), \quad t_1 \leq t \leq \hat{t}_2,$$

exists by assumption (the condition of unlimited extendability of solutions), and it follows from Theorem 4.4 on the continuous dependence of solutions and from Assertion 6.2 that the solutions

$$x^{(i)}(t) = x(t; \mu_\theta^{(i)}), \quad t_1 \leq t \leq \hat{t}_2,$$

converge uniformly to $x(t)$ as $i \to \infty$.

If $\mu_t \notin \mathfrak{M}_{x_2}$, then there exists an $\varepsilon > 0$ such that

$$|x(t) - x_2| \geq \varepsilon > 0 \quad \forall t \in [t_1, \hat{t}_2].$$

However, this is impossible, since every $x^{(i)}(t)$ takes on the value x_2 at some $t \in [t_1, \hat{t}_2]$. This proves that the set $\mathfrak{M}_{x_2}$ is closed.

The Existence of Optimal Solutions 149

It is also easy to prove that the function (8.7) is semicontinuous. Indeed, let

$$\mu_t^{(i)} \to \hat{\mu}_t \qquad (i \to \infty).$$

Then the trajectories

$$x^{(i)}(t) = x(t; \mu_\theta^{(i)}), \qquad t_1 \leq t \leq \hat{t}_2,$$

converge uniformly to the trajectory $\hat{x}(t) = x(t; \hat{\mu}_\theta)$ on the interval $[t_1, \hat{t}_2]$. Let

$$\hat{t} = \tau(\hat{\mu}_t), \qquad \hat{\mu}_t \in \mathfrak{M}_{x_2},$$

be the smallest value of t in $[t_1, t_2]$ for which the curve $\hat{x}(t) = x(t; \hat{\mu}_\theta)$, $t_1 \leq t \leq \hat{t}_2$, intersects the interval $x(t) \equiv x_2$, $t_1 \leq t \leq \hat{t}_2$. Then, the first instant of time at which $x^{(i)}(t)$ intersects the interval $x(t) \equiv x_2$ cannot be smaller than $\hat{t} - \varepsilon$ for i sufficiently large and any fixed $\varepsilon > 0$.

We know that the set $\mathfrak{M}_{x_2}$ is compact with respect to the weak convergence of generalized controls and that the function (8.7) is lower semicontinuous with respect to this convergence. From this, we can derive in a standard way that there exists a $\tilde{\mu}_t \in \mathfrak{M}_{x_2}$ such that

$$\tau(\tilde{\mu}_t) \leq \tau(\mu_t) \qquad \forall \mu_t \in \mathfrak{M}_{x_2}.$$

Indeed, we take a *minimizing sequence* of generalized controls $\mu_t^{(i)} \in \mathfrak{M}_{x_2}$, $i = 1, 2, \ldots$:

$$\tau(\mu_t^{(i)}) \to \inf_{\mu_t \in \mathfrak{M}_{x_2}} \tau(\mu_t) = \tilde{\tau} \geq t_1 \qquad (i \to \infty).$$

From this sequence, we choose a subsequence

$$\mu_t^{(i_1)}, \ldots, \mu_t^{(i_k)}, \ldots$$

which converges to a generalized control $\tilde{\mu}_t \in \mathfrak{M}_{x_2}$ because $\mathfrak{M}_{x_2}$ is compact. The lower semicontinuity of the function $\tau(\mu_t)$ yields

$$\lim_{k \to \infty} \inf \tau(\mu_t^{(i_k)}) \geq \tau(\tilde{\mu}_t).$$

By construction,

$$\lim_{k \to \infty} \inf \tau(\mu_t^{(i_k)}) = \lim_{k \to \infty} \tau(\mu_t^{(i_k)}) = \inf_{\mu_t \in \mathfrak{M}_{x_2}} \tau(\mu_t).$$

Therefore,

$$\tau(\tilde{\mu}_t) = \inf_{\mu_t \in \mathfrak{M}_{x_2}} \tau(\mu_t) = \tilde{\tau}.$$

This completes the proof of the theorem.

Remark. The condition of unlimited extendability of solutions guarantees the existence of the mapping

$$\mu_t \to x(t; \mu_\theta), \quad t_1 \leq t \leq t_2,$$

for any interval $[t_1, t_2]$. Without this condition the theorem is false.

It can happen without this condition that the trajectories

$$x^{(i)}(t) = x(t; \mu_\theta^{(i)}), \quad t_1 \leq t \leq t^{(i)}, \quad x^{(i)}(t_1) = x_1, \quad x^{(i)}(t^{(i)}) = x_2$$

are defined for a sequence of generalized controls $\mu_t^{(i)}$, where $t^{(1)} \geq t^{(2)} \geq \cdots$ and $t^{(i)} - t_1 \to$ infimum $(i \to \infty)$, yet the trajectory of the limit control

$$x(t; \mu_\theta), \quad x(t_1; \mu_\theta) = x_1$$

cannot be extended to the entire interval $t_1 \leq t \leq t_1 +$ infimum, since it goes to infinity.

8.3. The Existence Theorem in the Class of Ordinary Controls

The compactness of the set $U \subset R^r$ alone is not enough in order to guarantee the existence of a solution of the optimal problem in the class of usual controls,

$$\dot{x} = f(t, x, u), \quad u(t) \in \Omega_U,$$

$$t = t_1, \quad x(t_1) = x_1, \quad x(t_2) = x_2, \quad t_2 - t_1 \to \min. \tag{8.8}$$

(In this connection, we always assume, of course, that there exists at least one solution of the corresponding control problem

$$\dot{x} = f(t, x, u), \quad u(t) \in \Omega_U, \quad t = t_1, \quad x(t_1) = x_1, \quad x(t_2) = x_2, \tag{8.9}$$

and that the equation has the property of unlimited extendability of solutions.)

The reason for this is that, even in the case of a compact U, the set Ω_U is not compact in the topology of weak convergence of generalized controls. On the other hand, if we close Ω_U in this topology, then the closure will coincide (by the approximation lemma) with the set of all generalized controls $\mathfrak{M}_U$.

We shall present below a simple, yet typical, example of an optimal problem of the form (8.8) that does not have a solution. Now, we shall impose an additional condition on the right-hand side of equation (8.8). This condition consists of the requirement that

$$P(t, x) = \{f(t, x, u): u \in U\} = \operatorname{conv} P(t, x) \quad \forall (t, x) \in R \times R^n,$$

The Existence of Optimal Solutions

i.e., that the set $P(t, x)$ be convex for all (t, x). Under these assumptions, we shall prove Theorem 8.3 on the existence of a solution of the problem (8.8) (A.F. Filippov's theorem). The proof follows as a corollary from Theorem 8.2 and from Assertion 8.2 (the Filippov lemma) presented below.

Theorem 8.3. Assume that the optimal problem (8.8) has the property of unlimited extendability of solutions, and assume that there exists at least one solution

$$\hat{u}(t), \hat{x}(t), \quad t_1 \leq t \leq \hat{t}_2,$$

of the control problem (8.9). Further, assume that the set of admissible controls $U \subset R^r$ is compact, and that the right-hand side $f(t, x, u)$ satisfies the condition

$$P(t, x) = \{f(t, x, u): u \in U\} = \operatorname{conv} P(t, x) \quad \forall (t, x) \in R^{1+n}.$$

Then there exists an optimal solution

$$\tilde{u}(t), \tilde{x}(t), \quad t_1 \leq t \leq t_2,$$

of the control problem (8.9), i.e., a solution of the optimal problem (8.8).

Proof. By Theorem 8.2, there exists a solution

$$\tilde{\mu}_t, \tilde{x}(t), \quad t_1 \leq t \leq t_2,$$

of the corresponding convex optimal problem. Therefore, for almost all $t \in [t_1, t_2]$,

$$\dot{\tilde{x}}(t) = \langle \tilde{\mu}_t, f(t, \tilde{x}(t), u) \rangle, \tag{8.10}$$

and $t_2 - t_1$ is the minimal transfer time under the given boundary conditions in the class of generalized controls $\mathfrak{M}_U$.

We shall construct an admissible control $\tilde{u}(t) \in \Omega_U$ such that the trajectory $\tilde{x}(t)$, $t_1 \leq t \leq t_2$, constructed with the aid of the generalized control $\tilde{\mu}_t$, will also satisfy the equation

$$\dot{\tilde{x}}(t) = f(t, \tilde{x}(t), \tilde{u}(t))$$

for almost all $t \in [t_1, t_2]$, so that we shall obtain a solution

$$\tilde{u}(t), \tilde{x}(t), \quad t_1 \leq t \leq t_2,$$

of the optimal problem (8.8).

We use the notation

$$F(t) = \dot{x}(t) = \langle \tilde{\mu}_t, f(t, \tilde{x}(t), u) \rangle, \qquad t_1 \leq t \leq t_2.$$

It follows from the condition

$$P(t, x) = \operatorname{conv} P(t, x)$$

and from Assertion 2.1 that

$$F(t) \in P(t, \tilde{x}(t)) = \{ f(t, \tilde{x}(t), u) : u \in U \}.$$

Therefore, we can always choose a function $\tilde{u}(t)$, $t_1 \leq t \leq t_2$, which takes on values in U and satisfies the condition

$$F(t) = \dot{x}(t) = f(t, \tilde{x}(t), \tilde{u}(t)).$$

The basic difficulty lies in the demand that a function so chosen be measurable on $t_1 \leq t \leq t_2$. If we succeed in achieving this property, then, extending the function $\tilde{u}(t)$ outside of the interval $[t_1, t_2]$ in any admissible way, we obtain the required optimal control $\tilde{u}(t) \in \Omega_U$.

Using the notation

$$f(t, u) = f(t, \tilde{x}(t), u) - F(t), \qquad (t, u) \in [t_1, t_2] \times U,$$

we reduce the problem of finding a measurable $\tilde{u}(t), t_1 \leq t \leq t_2$, to the following assertion:

Assertion 8.2. Let $f(t, u)$ be an n-dimensional function defined for the values

$$(t, u) \in I \times U,$$

where I is an interval of the t-axis and U is a compact subset of R^r. Assume that $f(t, u)$ is (Lebesgue) measurable in t for a fixed $u \in U$ and continuous in u for a fixed $t \in I$.

Furthermore, assume that for each t in I the equation for u

$$f(t, u) = 0, \qquad u \in U, \tag{8.11}$$

has at least one solution. Then there exists a (Lebesgue) measurable function $u(t), t \in I$, which takes on its values in U and satisfies the equation

$$f(t, u(t)) = 0 \qquad \forall t \in I.$$

Proof. We denote by U_t, $t \in I$, the set of all solutions of equation (8.11) for the t under consideration. By assumption, the set U_t is nonempty

The Existence of Optimal Solutions

and compact for each $t \in I$ [since $f(t, u)$ is continuous in u and U is compact].

First of all, we shall prove that, for any compact subset $K \subset R^r$, the set

$$T = \{t: U_t \cap K \neq \emptyset\} \subset I \tag{8.12}$$

is (Lebesgue) measurable. We choose in U an everywhere dense sequence of points $u_1, u_2, \ldots$. We denote by V_ε the ε-neighborhood of the set $K \subset R^r$ and denote by W_ε the ε-neighborhood of the origin of R^n [the space of values of $f(t, u)$].

Since the functions $f(t, u_i)$, $t \in I$, $i = 1, 2, \ldots$, are measurable by assumption, the sets

$$T_{i,\varepsilon} = \{t: f(t, u_i) \in W_\varepsilon\} = [f(\cdot, u_i)]^{-1}(W_\varepsilon) \subset I$$

are measurable. Taking the union of the sets $T_{i,\varepsilon}$ over all i for which $u_i \in V_\varepsilon$ (ε is fixed), we obtain the measurable set

$$T_\varepsilon = \bigcup_i \{T_{i,\varepsilon}: u_i \in V_\varepsilon\}.$$

Obviously, $T_{\varepsilon'} \subset T_{\varepsilon''}$ for $\varepsilon' \leq \varepsilon''$. Therefore, the intersection of all T_ε, $\varepsilon > 0$, is measurable, since it can be represented as the intersection of the sequence of sets T_{ε_i}, where $\varepsilon_i \to 0$ ($i \to \infty$).

It is not difficult to see that $T \subset T_\varepsilon$ for all $\varepsilon > 0$. Indeed, let $\hat{t} \in T$. We choose an arbitrary point $\hat{u} \in U_{\hat{t}} \cap K$ and a subsequence $\{u_{i_k}\}$, $k = 1, 2, \ldots$, which converges to $\hat{u}$ as $k \to \infty$. Then by virtue of the continuity of $f(t, u)$ in u, we have:

$$f(\hat{t}, u_{i_k}) \to f(\hat{t}, \hat{u}) = 0 \quad (k \to \infty).$$

Therefore,

$$u_{i_k} \in V_\varepsilon, \quad f(\hat{t}, u_{i_k}) \in W_\varepsilon,$$

for k sufficiently large, i.e., $\hat{t} \in T_\varepsilon$. This proves the inclusion

$$T \subset \bigcap_{\varepsilon > 0} T_\varepsilon.$$

The opposite inclusion

$$T \supset \bigcap_{\varepsilon > 0} T_\varepsilon$$

can be also easily verified. Namely if a $\hat{t}$ belongs to the intersection, then, for every $\varepsilon > 0$, there exists a $u_{i_\varepsilon} \in V_\varepsilon$ such that

$$f(\hat{t}, u_{i_\varepsilon}) \in W_\varepsilon.$$

Therefore, one can choose from the points u_{i_ε} a sequence which converges to a point $\hat{u} \in K$. It follows from the continuity of $f(t, u)$ in u that $f(\hat{t}, \hat{u}) = 0$, i.e., $\hat{u} \in U_{\hat{t}} \cap K \neq \emptyset$, so that $\hat{t} \in T$. Thus, we have obtained the equality

$$T = \bigcap_{\varepsilon > 0} T_\varepsilon,$$

which proves that the set T is measurable.

After this basic preliminary fact has been established, we can pass directly to the construction of the required function $u(t)$, $t \in I$. To this end, we choose a value $a > 0$ so large that the r-dimensional cube

$$Q = \left\{ u = \begin{pmatrix} u^1 \\ u^r \end{pmatrix} : -a \leq u^i \leq a, i = 1, \ldots, r \right\} \subset R^r$$

contains the compact set U.

The partition of the interval $-a \leq \theta \leq a$ into 2^i subintervals of the same length $2a/2^i = a/2^{i-1}$,

$$I_j^{(i)} = \left\{ \theta : -a + (j-1)\frac{a}{2^{i-1}} \leq \theta \leq -a + j\frac{a}{2^{i-1}} \right\}, \qquad j = 1, 2, \ldots, 2^i,$$

will be called the partition of ith order. The Cartesian product of r intervals $I_{j_1}^{(i)}, \ldots, I_{j_r}^{(i)}$ which belong to the partition of order i,

$$I_{j_1}^{(i)} \times \cdots \times I_{j_r}^{(i)} \subset Q$$

will be called a cube of order i. Obviously, the number of such cubes is 2^{ri}.

Each cube of the $(i-1)$th order can be represented as the union of ith-order cubes. In particular, for each index $i = 1, 2, \ldots$, the union of all cubes of order i gives Q. For every $i = 1, 2, \ldots$, we number all cubes of order i in the sequence

$$Q_1^{(i)}, \ldots, Q_{2^{ri}}^{(i)}$$

in such a way that all cubes of order i contained in a given cube $Q_j^{(i-1)}$ of order $(i-1)$ have smaller indices than the cubes of order i contained in any cube $Q_k^{(i-1)}$ of order $(i-1)$ if the index k is larger than the given index j. In other words, the relations

$$Q_{k'}^{(i)} \subset Q_{l'}^{(i-1)}, \qquad Q_{k''}^{(i)} \subset Q_{l''}^{(i-1)}, \qquad l' < l'',$$

should imply the inequality $k' < k''$. Making use of this indexing, we construct, the following sequence of measurable functions,

$$u^{(i)}(t), \qquad t \in I, \qquad i = 1, 2, \ldots,$$

The Existence of Optimal Solutions

which take on values in U. Inside each cube $Q_j^{(i)}$, we choose a point

$$u_j^{(i)} \in Q_j^{(i)}, \quad j=1, 2, \ldots, 2^{ri},$$

and use the notation

$$T_j^{(i)} = \{t: U_t \cap Q_j^{(i)} \neq \emptyset\}, \quad j=1, 2, \ldots, 2^{ri}, i=1, 2, \ldots.$$

It follows from the measurability of the sets of form (8.12) that all sets $T_j^{(i)}$ are measurable. Moreover,

$$\sum_{j=1}^{2^{ri}} T_j^{(i)} = I, \quad i=1, 2, \ldots,$$

because the set U_t is not empty for all $t \in I$ and

$$\bigcup_{j=1}^{2^{ri}} Q_j^{(i)} = Q \supset U \supset U_t.$$

We set

$$u^{(i)}(t) = u_1^{(i)} \quad \forall t \in T_1^{(i)},$$

$$u^{(i)}(t) = u_k^{(i)} \quad \forall t = T_k^{(i)} \setminus \bigcup_{j=1}^{k-1} T_j^{(i)}, \quad k=2, \ldots, 2^{ri}, \quad i=1, 2, \ldots,$$

i.e., we set the function $u^{(i)}(t)$, $t \in I$, equal to $u_k^{(i)}$ at t, where k is the smallest of the indices of those ith-order cubes which have a nonempty intersection with U_t.

It is clear from the construction that the functions $u^{(i)}(t), t \in I$, are measurable and take on a finite number of values (no more than 2^{ri}).

We shall show that, for any $t \in I$, $i=1, 2, \ldots$, the relation $u^{(i)}(t) \in Q_k^{(i)}$ implies $u^{(i+p)}(t) \in Q_k^{(i)}$ for all $p = 1, 2, \ldots$. Let $u^{(i+p)}(t) \in Q_l^{(i+p)}$. Since $U_t \cap Q_k^{(i)} \neq \emptyset$, we can choose among the $(i+p)$th order cubes which are contained in $Q_k^{(i)}$ the cube $Q_{l'}^{(i+p)}$ with the smallest index l' which has a nonempty intersection with U_t. Therefore, $l \leq l'$. If the strict inequality $l < l'$ holds, then it means that the cube $Q_l^{(i+p)}$ is contained in a cube $Q_{k'}^{(i)}$ with a smaller index $k' < k$ than that of the cube $Q_k^{(i)}$. This means that $U_t \cap Q_{k'}^{(i)} \neq \emptyset$, which contradicts the method of constructing the functions $u^{(i)}(t)$.

Thus, the sequence of measurable functions $u^{(i)}(t)$ converges pointwise to a function $u(t)$ which takes on values in U, since the diameters of the cubes $Q_j^{(i)}$ tend to zero as $i \to \infty$, and since the set U is compact.

Therefore, the limit function $u(t)$ is measurable and, as it is not difficult to see, it satisfies the equation

$$f(t, u(t)) = 0 \quad \forall t \in I.$$

Indeed, since $u^{(i)}(t) \in Q_j^{(i)}$ for some j and $U_t \cap Q_j^{(i)} \neq \emptyset$, the distance from $u^{(i)}(t)$ to the closed set U_t is no larger than the diameter of $Q_j^{(i)}$, which tends to zero. Therefore, $\lim_{i \to \infty} u^{(i)}(t) = u(t) \in U_t$. This completes the proof of the assertion and, at the same time, of Theorem 8.3.

Remark. If a control problem has the form

$$\dot{x} = f(t, x) + B(t, x)u, \qquad u(t) \in \Omega_U, \qquad t = t_1, \qquad x(t_1) = x_1, \qquad x(t_2) = x_2,$$

where U is a convex compact subset of R^r, and the functions $f(t, x)$ and $B(t, x)$ are measurable in t and continuously differentiable with respect to x, then it follows from the theorem just proved that this problem has an optimal solution whenever there is at least one solution of the boundary problem and the right-hand side has the property of unlimited extendability of solutions.

Indeed, in this case the set

$$P(t, x) = \{ f(t, x) + B(t, x)u : u \in U \}$$

is convex.

8.4. Sliding Optimal Regimes

We begin with a simple example of the control problem (8.9) that satisfies all the conditions of Theorem 8.3 except the convexity condition and does not have an optimal solution.

We consider the following two-dimensional autonomous control problem (x, y, and u are scalars):

$$\begin{aligned} \dot{x} &= 1 - y^2, \\ \dot{y} &= u, \\ u(t) &\in U = \{1, -1\}, \\ x(0) &= y(0) = 0, \\ x(\tau) &= 1, \qquad y(\tau) = 0 \qquad (\tau > 0). \end{aligned} \qquad (8.13)$$

The set of all possible phase velocities depends only on y and consists of the two distinct vectors

$$\begin{pmatrix} 1 - y^2 \\ 1 \end{pmatrix} \quad \text{and} \quad \begin{pmatrix} 1 - y^2 \\ -1 \end{pmatrix}$$

in the $(\dot{x}, \dot{y})$-plane. Therefore, this set is not convex.

The Existence of Optimal Solutions

Obviously, we have $\tau > 1$ for any solution of the problem. It is also easy to see that there exist solutions with transfer time τ arbitrarily close to 1.

Consider a sequence of controls $u^{(i)}(t)$ which converges weakly to the generalized control

$$\tfrac{1}{2}\delta_1 + \tfrac{1}{2}\delta_{-1},$$

where each of the controls $u^{(i)}(t)$ rapidly oscillates between ± 1. To this sequence, there corresponds the sequence of trajectories

$$x^{(i)}(t), \; y^{(i)}(t), \qquad t \geq 0,$$

which converges uniformly on any interval $0 \leq t \leq \tau$ to the limit curve

$$x(t) = t, \qquad y(t) \equiv 0, \qquad t \geq 0.$$

Obviously, this curve is also a trajectory that corresponds to the generalized control $\tfrac{1}{2}\delta_1 + \tfrac{1}{2}\delta_{-1}$. The time of transfer from (0, 0) to (1, 0) along this curve is equal to 1, i.e., to the lower (not attainable) limit of the transfer time for the problem (8.13). For this reason, the motion ("sliding") of the phase point along the limit curve is called a sliding optimal regime of the control problem (8.13).

We can conclude on the basis of Theorem 8.1 that this situation is a general one in the case of a compact U. Thus, every *sliding regime*, i.e., the uniform limit of trajectories of the problem

$$\dot{x} = f(t, x, u), \quad u(t) \in \Omega_U, \quad t = t_1, \quad x(t_1) = x_1, \quad x(t_2) = x_2,$$

can be realized as a trajectory that corresponds to a generalized control.

It turns out that an even stronger assertion holds. Namely, every sliding regime is a trajectory that corresponds to a chattering control, i.e., to a generalized control of the form [see formula (2.6)]

$$\mu_t = \sum_{i=1}^{p} \mu_i(t) \delta u_i(t), \quad u_i(t) \in \Omega_U, \quad \mu_i(t) \geq 0, \quad \sum_{i=1}^{p} \mu_i(t) \equiv 1.$$

For this reason, such controls are also called *sliding* controls. An investigation of optimal regimes can be restricted to sliding controls only. This follows directly from the following assertion:

Assertion 8.3. If the convex control problem

$$\dot{x} = \langle \mu_t, f(t, x, u) \rangle, \quad \mu_t \in \mathfrak{M}_U, \quad t = t_1, \quad x(t_1) = x_1, \quad x(t_2) = x_2,$$

satisfies the assumptions of Theorem 8.2, then, for any solution $\hat{\mu}_t, \hat{x}(t)$,

$t_1 \leq t \leq t_2$, one can choose a chattering control

$$\lambda_t = \sum_{i=0}^{n} \lambda_i(t)\delta_{u_i(t)}, \quad u_i(t) \in \Omega_U, \quad \lambda_i(t) \geq 0, \quad \sum_{i=0}^{n} \lambda_i(t) \equiv 1$$

[n is the dimensionality of the column $f(t, x, u)$] such that the curve $\hat{x}(t)$ satisfies the equation

$$\dot{\hat{x}}(t) = \left\langle \sum_{i=0}^{n} \lambda_i(t)\delta_{u_i(t)}, f(t, \hat{x}(t), u) \right\rangle = \sum_{i=0}^{n} \lambda_i(t) f(t, \hat{x}(t), u_i(t)).$$

This means that $\hat{x}(t)$ is a trajectory of the problem under consideration that corresponds to the chattering control λ_t.

Proof. We shall make use of the following simple assertion, which is known as Carathéodory's lemma, and which we prove below:

The convex hull of an arbitrary set $M \subset R^m$ coincides with the union of the convex hulls of all sets of $m+1$ points of the set M. Let Λ denote the n-dimensional simplex

$$\Lambda = \left\{ \lambda = (\lambda_0, \lambda_1, \ldots, \lambda_n) : \lambda_i \geq 0, \sum_{i=0}^{n} \lambda_i = 1 \right\},$$

let W denote the $(n+1)$th Cartesian power of the set U,

$$W = \overbrace{U \times U \times \cdots \times U}^{n+1} = \{w = (u_0, \ldots, u_n) : u_i \in U, i = 0, \ldots, n\},$$

and let $g(t, \lambda, w)$ denote the function

$$g(t, \lambda, w) = \sum_{i=0}^{n} \lambda_i f(t, \hat{x}(t), u_i) - \langle \hat{\mu}_t, f(t, \hat{x}(t), u) \rangle,$$

$$t \in [t_1, t_2], \quad \lambda = (\lambda_0, \lambda_1, \ldots, \lambda_n) \in \Lambda, \quad w = (u_0, u_1, \ldots u_n) \in W.$$

Let $f(t, \hat{x}(t), u_i)$, $i = 0, 1, \ldots, n+1$, be $(n+1)$ arbitrary points in $P(t, \hat{x}(t))$. Then the convex hull of this set of $(n+1)$ points consists of all points of the form

$$\sum_{i=0}^{n} \lambda_i f(t, \hat{x}(t), u_i).$$

On the basis of Assertion 2.1, the vector $\langle \hat{\mu}_t, f(t, \hat{x}(t), u) \rangle$ belongs to the set conv $P(t, \hat{x}(t))$. Therefore, it follows from Carathéodory's lemma that, for

The Existence of Optimal Solutions 159

all $t \in [t_1, t_2]$, the equation for (λ, w)

$$g(t, \lambda, w) = 0, \qquad (\lambda, w) \in \Lambda \times W,$$

is solvable. Making use of Assertion 8.2, we can choose measurable functions

$$\lambda(t) \in \Lambda, \qquad w(t) \in W, \qquad t \in [t_1, t_2],$$

that satisfy the equation

$$g(t, \lambda(t), w(t)) = 0$$

identically with respect to $t \in [t_1, t_2]$. Therefore, we obtain the following equality for almost all $t \in [t_1, t_2]$:

$$\dot{x}(t) = \langle \hat{\mu}_t, f(t, \hat{x}(t), u) \rangle = \sum_{i=0}^{n} \lambda_i(t) f(t, \hat{x}(t), u_i(t)) = \left\langle \sum_{i=0}^{n} \lambda_i(t) \delta_{u_i(t)}, f(t, \hat{x}(t), u) \right\rangle,$$

which proves the assertion.

Proof of Carathéodory's Lemma. We shall show that, if

$$z = \sum_{i=0}^{p} \lambda_i z_i, \qquad z_i \in M, \qquad \lambda_i > 0, \qquad \sum_{i=0}^{p} \lambda_i = 1, \qquad p > m,$$

then the point z can be represented as a similar sum, but with a smaller number of points z_i, which immediately implies the lemma.

Since the vectors $z_1 - z_0, \ldots, z_p - z_0 \in R^m$ are linearly dependent ($p > m$), there exist values $v_1, \ldots, v_p$ which are not all zero and such that

$$\sum_{i=1}^{p} v_i(z_i - z_0) = -\sum_{i=1}^{p} v_i z_0 + \sum_{i=1}^{p} v_i z_i = 0.$$

Using the notation $v_0 = -\sum_{i=1}^{p} v_i$, we obtain

$$\sum_{i=0}^{p} v_i z_i = 0, \qquad \sum_{i=0}^{p} v_i = 0.$$

Obviously, an ε can be chosen so that all $\lambda_i + \varepsilon v_i$ are nonnegative for $i = 0, 1, \ldots, p$, and at least one of these numbers, say $\lambda_j + \varepsilon v_j$, is zero,

$$\lambda_i + \varepsilon v_i \geq 0, \qquad i = 0, 1, \ldots, p, \qquad \lambda_j + \varepsilon v_j = 0.$$

Assume that the points are numbered so that $j = 0$ yields $\lambda_0 + \varepsilon v_0 = 0$.

Then we obtain the equalities

$$z = \sum_{i=0}^{p} (\lambda_i + \varepsilon v_i) z_i = \sum_{i=1}^{p} (\lambda_i + \varepsilon v_i) z_i, \qquad \lambda_i + \varepsilon v_i \geq 0, \; i = 1, \ldots, p,$$

$$\sum_{i=1}^{p} (\lambda_i + \varepsilon v_i) = \sum_{i=0}^{p} (\lambda_i + \varepsilon v_i) = \sum_{i=0}^{p} \lambda_i = 1,$$

which prove the lemma.

With the aid of the procedure for constructing optimal regimes that we described here, the theorem on the existence of optimal solutions in the class of ordinary controls can be given a form more general than Theorem 8.3. This generalization is often useful (see Theorem 8.4), and it is given by the following assertion:

Assertion 8.4. Let all the conditions of Theorem 8.3 except the condition

$$P(t, x) = \{ f(t, x, u) : u \in U \} = \operatorname{conv} P(t, x)$$

be satisfied, and let this condition be replaced by a more general assumption: The set $P(t, x)$ contains the boundary (in particular, coincides with the boundary) of its convex hull

$$\operatorname{conv} P(t, x) \subset R^n.$$

Then there exists an optimal solution

$$\tilde{u}(t), \tilde{x}(t), \qquad t_1 \leq t \leq t_2,$$

of the control problem (8.9), i.e., a solution of the optimal problem (8.8).

Proof. On the basis of Theorem 8.2 and Assertion 8.3, there exists an optimal solution

$$\tilde{\mu}_t, \tilde{x}(t), \qquad t_1 \leq t \leq t_2,$$

where $\tilde{\mu}_t$ is a chattering control,

$$\tilde{\mu}_t = \tilde{\lambda}_t = \sum_{i=0}^{n} \tilde{\lambda}_i(t) \delta_{\tilde{u}_i}(t), \qquad \tilde{u}_i(t) \in \Omega u, \qquad \tilde{\lambda}_i(t) \geq 0, \qquad \sum_{i=0}^{n} \tilde{\lambda}_i(t) \equiv 1.$$

Since

$$(\tilde{\lambda}(t); \tilde{w}(t)) = (\tilde{\lambda}_0(t), \ldots, \tilde{\lambda}_n(t); \tilde{u}_0(t), \ldots, \tilde{u}_n(t)),$$

$$\tilde{x}(t), \qquad t_1 \leq t \leq t_2,$$

The Existence of Optimal Solutions

is an optimal solution of the control problem

$$\dot{x} = \sum_{i=0}^{n} \lambda_i f(t, x, u_i), \quad (\lambda(t); w(t)) \in \Omega_{\Lambda \times W},$$

$$t = t_1, \quad x(t_1) = x_1, \quad x(t_2) = x_2,$$

there exists, in accordance with the maximum principle, an absolutely continuous function $\tilde{\psi}(t) \neq 0$, $t_1 \leq t \leq t_2$, such that, for almost all $t \in [t_1, t_2]$,

$$\tilde{\psi}(t) \sum_{i=0}^{n} \tilde{\lambda}_i(t) f(t, \tilde{x}(t), \tilde{u}_i(t)) = \max_{(\lambda, w) \in \Lambda \times W} \tilde{\psi}(t) \sum_{i=0}^{n} \lambda_i f(t, \tilde{x}(t), u_i),$$

$$(\lambda, w) = (\lambda_0, \ldots, \lambda_n; u_0, \ldots, u_n).$$

This means that, for almost all $t \in [t_1, t_2]$, the phase velocity vector

$$\dot{\tilde{x}}(t) = \sum_{i=0}^{n} \tilde{\lambda}_i(t) f(t, \tilde{x}(t), \tilde{u}_i(t))$$

belongs to the boundary of the set conv $P(t, \tilde{x}(t))$ and, therefore, also to the set $P(t, \tilde{x}(t))$. By Assertion 8.2, it follows that we can choose an admissible control

$$\tilde{u}(t) \in \Omega_U$$

such that

$$f(t, \tilde{x}(t), \tilde{u}(t)) = \dot{\tilde{x}}(t).$$

Therefore,

$$\tilde{u}(t), \tilde{x}(t), \quad t_1 \leq t \leq t_2,$$

is a required solution.

Remark. If the function $f(t, x, u)$ does not depend on time,

$$f(t, x, u) = f(x, u)$$

(see Section 8.5), then Assertion 8.4 can be proved without using the maximum principle.

Obviously, it is sufficient to show that the set T of all points t of the interval $[t_1, t_2]$ at which the phase velocity vector

$$\sum_{i=0}^{n} \tilde{\lambda}_i(t) f(\tilde{x}(t), \tilde{u}_i(t)) = f(t)$$

does not belong to the boundary of the set conv $P(\tilde{x}(t))$ has Lebesgue measure zero.

Let T_k be the set of instants of time $t \in [t_1, t_2]$ at which

$$(1 + 1/k) f(t) \in \text{conv } P(\tilde{x}(t)).$$

It is easy to see that $T = \bigcup_{k=1}^{\infty} T_k$. Therefore, if meas $T > 0$, then meas $T_{\hat{k}} > 0$ for some $\hat{k}$.

We shall show that, if such a $\hat{k}$ exists, then the time $t_2 - t_1$ of transfer from $\tilde{x}(t_1)$ to $\tilde{x}(t_2)$ can be decreased, which contradicts the assumption of the optimality of the solution $\bar{\mu}_t$, $\tilde{x}(t)$, $t_1 \leq t \leq t_2$.

We denote by $\omega(t)$, $t_1 \leq t \leq t_2$, the function which is equal to 1 for $t \notin T_{\hat{k}}$ and equal to $1 + 1/\hat{k}$ for $t \in T_{\hat{k}}$. We introduce the new time τ with the aid of the equation

$$\frac{d\tau}{dt} = \frac{1}{\omega(t)}, \quad \tau(t_1) = 0.$$

When t changes from t_1 up to t_2, the time τ changes from 0 up to

$$\hat{\tau} = \int_{t_1}^{t_2} \frac{dt}{\omega(t)} = t_2 - t_1 - \text{meas } T_{\hat{k}} + \frac{\text{meas } T_{\hat{k}}}{1 + 1/\hat{k}} < t_2 - t_1.$$

The functions

$$\bar{\mu}_{t(\tau)} \quad \text{and} \quad \tilde{x}(t(\tau)), \quad 0 \leq \tau \leq \hat{\tau},$$

satisfy the equation

$$\frac{d\tilde{x}}{d\tau} = \omega(t(\tau)) \sum_{i=0}^{n} \tilde{\lambda}_i(t(\tau)) f(\tilde{x}(t(\tau)), u_i(t(\tau))) \in \text{conv } P(\tilde{x}(t(\tau)),$$

and the boundary conditions

$$\tilde{x}(t(0)) = \tilde{x}(t_1), \quad \tilde{x}(t(\hat{\tau})) = \tilde{x}(t_2).$$

Therefore, they do improve the transfer time $t_2 - t_1$.

8.5. The Existence Theorem for Regular Problems of the Calculus of Variations

As an illustration of the results obtained, we shall prove a classical theorem of the calculus of variations on the existence of a solution for a regular problem of the calculus of variations. Essential parts of this theorem belong to Hilbert and Tonelli.

The Existence of Optimal Solutions

Let

$$f(\tau, z, u) \quad \text{with} \quad (\tau, z, u) \in R \times R^n \times R^n$$

be a continuously differentiable function of its arguments. We shall say that this function is convex in the argument u if, for arbitrary fixed τ and z,

$$f(\tau, z, \alpha u_1 + \beta u_2) \leq \alpha f(\tau, z, u_1) + \beta f(\tau, z, u_2)$$
$$\forall \alpha, \beta \geq 0, \quad \alpha + \beta = 1, \quad \text{and} \quad \forall u_1, u_2 \in R^n.$$

The *simple problem* of the calculus of variations consists of finding the minimum of the integral

$$J(z(\tau)) = \int_{\tau_1}^{\tau_2} f\left(\tau, z(\tau), \frac{dz(\tau)}{d\tau}\right) d\tau$$

in the class of all absolutely continuous functions

$$z(\tau), \quad \tau_1 \leq \tau \leq \tau_2,$$

that satisfy given boundary conditions

$$z(\tau_1) = z_1, \quad z(\tau_2) = z_2.$$

The variational problem

$$\int_{\tau_1}^{\tau_2} f\left(\tau, z(\tau), \frac{dz(\tau)}{d\tau}\right) d\tau \to \min$$

is said to be *regular* if the function $f(\tau, z, u)$ is positive for all values of its arguments and convex in u.*

Theorem 8.4. The simple problem of the calculus of variations

$$\int_{\tau_1}^{\tau_2} f\left(\tau, z, \frac{dz}{d\tau}\right) d\tau \to \min, \quad z(\tau_1) = z_1, \quad z(\tau_2) = z_2,$$

has a solution $z(\tau)$, $\tau_1 \leq \tau \leq \tau_2$, if the problem is regular and satisfies the

---

*In this definition, the requirement of being positive can be replaced by the requirement that the function be bounded from below, since two variational problems with fixed end points are equivalent if the difference of the integrands is constant.

additional condition

$$\frac{|u|}{f(\tau, z, u)} \to 0 \quad (|u| \to \infty), \tag{8.14}$$

where τ and z are arbitrary and fixed.

The proof consists of a conceptually simple reduction of the simple problem of the calculus of variations to a time-optimal problem with fixed end points, and of the subsequent application of Assertion 8.4. This reduction turns out to be possible because the indefinite integral

$$t = \int_{\tau_1}^{\tau} f\left(\theta, z(\theta), \frac{dz(\theta)}{d\theta}\right) d\theta, \quad \tau \in [\tau', \tau''],$$

along an absolutely continuous curve $z(\tau)$, $\tau' \leq \tau \leq \tau''$, is a strictly monotonic function of $\tau (dt/d\tau = f(\tau, z, dz/d\tau) > 0)$. Therefore, taking this integral as a new independent variable, time, and considering τ and z as phase variables and the derivative $dz/d\tau$ as the control parameter u, we can write an equivalent optimal problem in the form

$$\frac{d\tau}{dt} = \frac{1}{f(\tau, z, u)}, \quad \frac{dz}{dt} = \frac{dz}{d\tau}\frac{d\tau}{dt} = \frac{u}{f(\tau, z, u)},$$

$$\tau(0) = \tau_1, \quad \tau(T) = \tau_2, \quad z(0) = z_1, \quad z(T) = z_2, \quad T \to \min.$$

Thus, we consider $(\tau, x) \in R \times R^n$ as the phase point in R^{1+n} and $u \in R^n$ as the control parameter, and we write the following autonomous control problem:

$$\frac{d\tau}{dt} = \frac{1}{f(\tau, x, u)}, \quad \frac{dx}{dt} = \frac{u}{f(\tau, x, u)}, \quad u(t) \in \Omega_{R^n},$$

$$\tau(0) = \tau_1, \quad \tau(T) = \tau_2, \quad x(0) = z_1, \quad x(T) = z_2. \tag{8.15}$$

If we add the point $(0, 0) \in R \times R^n$ to the set of all possible phase velocities of the problem (8.15) for given τ and x, then, by virtue of the condition (8.14), this set becomes a compact subset of R^{1+n}. We denote this set by $P(\tau, x)$,

$$P(\tau, x) = \{(0, \mathbf{0})\} \cup \left\{ \left(\frac{1}{f(\tau, x, u)}, \frac{u}{f(\tau, x, u)} \right) : u \in R^n \right\} \subset R^{1+n}.$$

The set $P(\tau, x)$ can be represented as a homeomorphic image of an n-dimensional sphere $S^n \subset R^{n+1}$. To this end, we fix a point $\hat{w} \in S^n$ and first give a

The Existence of Optimal Solutions

homeomorphism (e.g., the stereographic projection)

$$u:(S^n\setminus\{\hat{w}\})\to R^n \tag{8.16}$$

between $(S^n\setminus\{\hat{w}\})$ and R^n under which an arbitrary sequence of points $w_i \in S^n\setminus\{\hat{w}\}$ that converges to $\hat{w}$ becomes a sequence of points $u(w_i)\in R^n$ converging to infinity. We consider the following functions of w, which are continuous on $S^n\setminus\{\hat{w}\}$:

$$\frac{1}{f(\tau, x, u(w))} \quad \text{and} \quad \frac{u(w)}{f(\tau, x, u(w))}.$$

Setting these functions equal to zero at $\hat{w}$, we can extend them, by the condition (8.14), to functions which are continuous on the entire sphere S^n.

It can be seen directly that the continuous mapping just defined is one-to-one and, therefore, homeomorphic, since S^n is compact.

Thus, we have defined the autonomous control problem

$$\frac{d\tau}{dt}=\frac{1}{f(t,x,u(w))}, \quad \frac{dx}{dt}=\frac{u(w)}{f(\tau,x,u(w))}, \quad w(t)\in\Omega_{S^n}, \tag{8.17}$$

$$\tau(0)=\tau_1, \quad \tau(T)=\tau_2, \quad x(0)=z_1, \quad x(T)=z_2.$$

If we add the requirement that the transfer time T be minimal,

$$T\to\min, \tag{8.18}$$

then this problem becomes an optimal problem of the form (1.3) with the compact set of admissible values $U=S^n\subset R^{n+1}$.

We consider an arbitrary solution of the control problem (8.17)

$$w(t), \tau(t), x(t), \quad 0\leqslant t\leqslant T, \tag{8.19}$$

that satisfies the following condition:

$$w(t)\neq\hat{w}$$

for almost all $t\in[0, T]$. We shall show that, to every absolutely continuous curve

$$z(\tau), \quad \tau_1\leqslant\tau\leqslant\tau_2, \quad z(\tau_1)=z_1, \quad z(\tau_2)=z_2, \tag{8.20}$$

one can assign a solution of the form (8.19) of the control problem (8.17) in a certain standard way, and that the transfer time T for the solution (8.19) will be equal to the integral $J(z(\tau))$,

$$T = J(z(\tau)) = \int_{\tau_1}^{\tau_2} f\left(\tau, z(\tau), \frac{dz(\tau)}{d\tau}\right) d\tau. \tag{8.21}$$

Moreover, it will turn out that every solution (8.19) can be obtained with the aid of this correspondence.

Therefore, in order to prove the theorem it will be sufficient to show that the optimal problem (8.17)–(8.18) has a solution of the form (8.19).

In order to pass from the absolutely continuous curve (8.20) to a solution of the form (8.19) that satisfies the relation (8.21), we introduce the absolutely continuous function

$$t = t(\tau) = \int_{\tau_1}^{\tau} f\left(\theta, z(\theta), \frac{dz(\theta)}{d\theta}\right) d\theta, \qquad \tau \in [\tau_1, \tau_2].$$

Since the derivative of this function is greater than zero for almost all $\tau \in [\tau_1, \tau_2]$,

$$\frac{dt}{d\tau} = f\left(\tau, z(\tau), \frac{dz(\tau)}{d\tau}\right) > 0,$$

the inverse function

$$\tau = \tau(t), \qquad 0 = t(\tau_1) \leqslant t \leqslant T = t(\tau_2) = \int_{\tau_1}^{\tau_2} f\left(\theta, z(\theta), \frac{dz(\theta)}{d\theta}\right) d\theta,$$

is also absolutely continuous.

We define the solution (8.19), setting

$$w(t) = u^{-1}\left(\frac{dz(\tau(t))}{d\tau}\right), \qquad \tau(t), \qquad x(t) = z(\tau(t)), \qquad 0 \leqslant t \leqslant T = J(z(\tau)),$$

where u^{-1} is the mapping inverse to (8.16). The functions

$$\tau(t) \quad \text{and} \quad x(t), \qquad 0 \leqslant t \leqslant T,$$

are absolutely continuous and satisfy the given boundary conditions [the absolute continuity of the composite function $x(t) = z(\tau(t))$ follows from the absolute continuity of the functions $z(\tau)$ and $\tau(t)$ and from the fact that $d\tau/dt > 0$ a.e.]. The function $w(t)$ is measurable because u^{-1} is a continuous function. The function $dz(\tau)/d\tau$ is measurable, and $\tau(t)$ is an absolutely con-

The Existence of Optimal Solutions

tinuous function whose derivative is greater than zero almost everywhere. After these remarks, one can convince oneself by direct differentiation that the functions $w(t)$, $\tau(t)$, and $x(t)$ satisfy the differential equations (8.17).

We shall now define the curve (8.20) corresponding to the solution of the form (8.19). One can return from this curve to (8.19) with the aid of the procedure described in the preceding paragraph.

The function

$$t(\tau), \quad \tau_1 \leq t \leq \tau_2,$$

inverse to the function $\tau(t)$, $0 \leq t \leq T$, is absolutely continuous, since $w(t) \neq \hat{w}$ for almost all $t \in [0, T]$, so that

$$\frac{d\tau}{dt} = \frac{1}{f(\tau(t), x(t), u(w(t)))} > 0.$$

We set

$$z(\tau) = x(t(\tau)), \quad \tau_1 \leq \tau \leq \tau_2.$$

The curve thus defined is absolutely continuous [because the functions $x(t)$ and $t(\tau)$ are absolutely continuous and the derivative of $t(\tau)$ is positive almost everywhere] and satisfies the boundary conditions

$$z(\tau_1) = z_1 \quad \text{and} \quad z(\tau_2) = z_2.$$

We shall show that, for almost all $t \in [0, T]$,

$$\frac{dz(\tau(t))}{d\tau} = u(w(t))$$

and

$$J(z(\tau)) = \int_{\tau_1}^{\tau_2} f\left(\tau, z(\tau), \frac{dz(\tau)}{d\tau}\right) d\tau = T.$$

The first equality can be verified by direct differentiation. One obtains the second equality by integrating the relation

$$\frac{dt}{d\tau} = f(\tau, x(t(\tau)), u(w(\tau)))$$

with respect to $d\tau$ from τ_1 to τ_2, where the functions $x(t(\tau))$ and $u(w(t(\tau)))$ have been replaced by the functions $z(\tau)$ and $dz(\tau)/d\tau$, respectively.

Thus, in order to prove the theorem it remains to prove that the optimal problem (8.17)–(8.18) has a solution of the form (8.19).

The existence of such a solution follows easily from Assertion 8.4 and the maximum principle if we show that, for any τ and x, the set $P(\tau, x) \subset R^{1+n}$ is the boundary of a bounded open convex set

$$D(\tau, x) \subset R^{1+n}.$$

Indeed, in this case the boundary of the convex hull conv $P(\tau, x)$ coincides with the set $P(\tau, x)$. Moreover, since the control problem (8.17) always has a solution, and since the system of equations (8.17) has the property of unlimited extendability of solutions, there exists, on the basis of Assertion 8.4, a solution of the optimal problem (8.17)–(8.18)

$$\tilde{w}(t), \tilde{\tau}(t), \tilde{x}(t), \qquad 0 \leqslant t \leqslant T.$$

Making use of the maximum principle, we shall prove that every optimal solution satisfies the condition

$$\tilde{w}(t) \neq \hat{w}$$

for almost all $t \in [0, T]$. Since the system under consideration is autonomous, there exists, by the maximum principle and Assertion 1.1, an absolutely continuous function

$$\tilde{\psi}(t) = (\tilde{\chi}_0(t), \chi(t)) \neq 0, \qquad 0 \leqslant t \leqslant T,$$

such that for almost all $t \in [0, T]$:

$$\frac{\tilde{\chi}_0(t)}{f(\tilde{\tau}(t), \tilde{x}(t), u(\tilde{w}(t)))} + \frac{\tilde{\chi}(t)u(\tilde{w}(t))}{f(\tilde{\tau}(t), \tilde{x}(t), u(\tilde{w}(t)))} = \sup_{u \in R^n} \frac{1}{f(\tilde{\tau}(t), \tilde{x}(t), u)} (\tilde{\chi}_0(t) + \tilde{\chi}(t)u)$$

$$= \text{const} \geqslant 0$$

If const > 0, then $\tilde{w}(t) \neq \hat{w}$ for almost all $t \in [0, T]$, because

$$\frac{1}{f(\tau, x, u(\hat{w}))} = \frac{u(\hat{w})}{f(\tau, x, u(\hat{w}))} = 0$$

at the point $\hat{w}$. If const $= 0$, then $\tilde{\chi}(t) \equiv 0$, $\tilde{\chi}_0(t) < 0$ and, therefore,

$$\tilde{w}(t) = \hat{w}$$

for almost all $t \in [0, T]$. But this is impossible if $\tau_1 \neq \tau_2$, because if $\tilde{w}(t) \equiv \hat{w}$ then the phase point $(\tilde{\tau}(t), \tilde{x}(t))$, $0 \leqslant t \leqslant T$, would be stationary.

Thus, in order to complete the proof of the theorem, it remains to construct a bounded open convex set $D(\tau, x)$. We shall make use of the para-

The Existence of Optimal Solutions

metric representation of the set

$$Q(\tau, x) = P(\tau, x) \setminus \{(0, 0)\}$$

in the form

$$Q(\tau, x) = \left\{ (v_0, v) = \left(\frac{1}{f(\tau, x, u)}, \frac{u}{f(\tau, x, u)} \right) : u \in R^n \right\} \subset R^{1+n},$$

i.e.,

$$v_0 = \frac{1}{f(\tau, x, u)}, \quad v = \frac{u}{f(\tau, x, u)}, \quad u \in R^n.$$

It follows from these equalities that $u = v/v_0$. Therefore, $Q(\tau, x)$ coincides with the set of all $(v_0, v) \in R^{1+n}$ that satisfy the relations

$$v_0 f(\tau, x, v/v_0) = 1, \quad v_0 > 0.$$

We define $D(\tau, x)$ as the set of points from the half-space $v_0 > 0$ in R^{1+n} that satisfy the inequality

$$v_0 f(\tau, x, v/v_0) < 1.$$

This set is open by virtue of the continuity of the function f. Let us prove that it is bounded. If $|(v_0, v)| \to \infty$, then either $|v_0| \to \infty$ or $|v| \to \infty$. For $|v_0| \to \infty$, we have

$$|v_0 f(\tau, x, v/v_0)| \to \infty,$$

since the function $f(\tau, x, u)$ with fixed τ and x is bounded from below by a positive constant [by virtue of continuity and by condition (8.14)]. If $|v| \to \infty$, then

$$|v_0 f(\tau, x, v/v_0)| = |v| \frac{f(\tau, x, v/v_0)}{|v/v_0|} \to \infty$$

[also by virtue of condition (8.14)].

Furthermore, it is easy to see that $P(\tau, x)$ contains the boundary of the set $D(\tau, x)$. Indeed, if $(v_0, v) \in D(\tau, x)$, where $(v_0, v) \to (\hat{v}_0, \hat{v})$, $\hat{v}_0 \neq 0$, then

$$v_0 f(\tau, x, v/v_0) \to \hat{v}_0 f(\tau, x, \hat{v}/\hat{v}_0) = 1;$$

and if $(\hat{v}_0, \hat{v}) \notin D(\tau, x)$, then

$$\hat{v}_0 f(\tau, x, \hat{v}/\hat{v}_0) = 1,$$

i.e., $(\hat{v}_0, \hat{v}) \in P(\tau, x)$. On the other hand, if $\hat{v}_0 = 0$, then also $\hat{v} = 0$. Indeed, in the case $\hat{v} \neq 0$ we can assume that $|v| \neq 0$ and, therefore,

$$v_0 f(\tau, x, v/v_0) = |v| \frac{f(\tau, x, v/v_0)}{|v|/v_0} \to \infty,$$

which contradicts the condition $(v_0, v) \in D(\tau, x)$.

Conversely, every point $(\hat{v}_0, \hat{v}) \in P(\tau, x)$ can be represented as a limit of points $(v_0, v) \in D(\tau, x)$, i.e., a limit of points (v_0, v) for which

$$v_0 f(\tau, x, v/v_0) < 1, \qquad v_0 > 0.$$

Indeed, if $\hat{v}_0 \neq 0$, then $\hat{v}_0 > 0$ and, for $0 < \lambda < 1$, we have $(\lambda \hat{v}_0, \lambda \hat{v}) \in D(\tau, x)$ because

$$\lambda \hat{v}_0 f(\tau, x, \lambda \hat{v}/\lambda \hat{v}_0) = \lambda \hat{v}_0 f(\tau, x, \hat{v}/\hat{v}_0) \leq \lambda < 1.$$

Since $(\lambda \hat{v}_0, \lambda \hat{v}) \to (\hat{v}_0, \hat{v})$ as $\lambda \to 1$, we have that $(\hat{v}_0, \hat{v})$ is a limit of points in $D(\tau, x)$. If $\hat{v}_0 = 0$ and, therefore, $\hat{v} = \mathbf{0}$, then we obtain as $v_0 \to 0$ $(v_0 > 0)$

$$(v_0, \mathbf{0}) \to (\hat{v}_0, \hat{v}) = (0, \mathbf{0}),$$

and, for a $v_0 > \mathbf{0}$ sufficiently small,

$$v_0 f(\tau, x, \mathbf{0}) < 1,$$

i.e., $(v_0, \mathbf{0}) \in D(\tau, x)$.

Finally, the convexity of the set $D(\tau, x)$ follows easily from the convexity of $f(\tau, x, u)$ in u. Setting

$$v'_0 f(\tau, x, v'/v'_0) < 1, \qquad v''_0 f(\tau, x, v''/v''_0) < 1, \qquad v'_0, v''_0 > 0,$$

$$\alpha, \beta \geq 0, \qquad \alpha + \beta = 1,$$

we shall show that

$$(\alpha v'_0 + \beta v''_0) f\left(\tau, x, \frac{\alpha v' + \beta v''}{\alpha v'_0 + \beta v''_0}\right) < 1.$$

Using the notation $\Delta = \alpha v'_0 + \beta v''_0$ and making use of the convexity of f, we can write

$$\Delta f\left(\tau, x, \frac{\alpha v' + \beta v''}{\Delta}\right) = \Delta f\left(\tau, x, \frac{\alpha v'_0}{\Delta} \frac{v'}{v'_0} + \frac{\beta v''_0}{\Delta} \frac{v''}{v''_0}\right)$$

$$\leq \Delta \left\{\frac{\alpha v'_0}{\Delta} f\left(\tau, x, \frac{v'}{v'_0}\right) + \frac{\beta v''_0}{\Delta} f\left(\tau, x, \frac{v''}{v''_0}\right)\right\}$$

$$= \alpha v'_0 f\left(\tau, x, \frac{v'}{v'_0}\right) + \beta v''_0 f\left(\tau, x, \frac{v''}{v''_0}\right) < \alpha + \beta = 1.$$

Bibliography

The reader will find a detailed motivation of the mathematical formalism used in the present work in the following book:

> L. S. PONTRYAGIN, V. G. BOLTYANSKII, R. V. GAMKRELIDZE, and E. F. MISHCHENKO, *The Mathematical Theory of Optimal Processes*, J. Wiley, New York, 1962.

A good modern presentation of the theory of optimal processes is contained in the following books, which complement each other:

> H. HERMES and J. P. LASALLE, *Functional Analysis and Time-Optimal Control*, Academic Press, New York, 1969;
>
> L. D. BERKOVITZ, *Optimal Control Theory* (Applied Mathematical Sciences, Vol. 12), Springer-Verlag, New York, 1974.
>
> J. WARGA, *Optimal Control of Differential and Functional Equations*, Academic Press, New York, 1972.

Warga's book contains a detailed presentation of all the auxiliary material, including measure theory and the theory of convex sets.

Of the books devoted specifically to the presentation of the theory of convex sets and functions and the theory of Radon measures, we suggest the following:

> R. T. ROCKAFELLAR, *Convex Analysis*, Princeton University Press, Princeton, New Jersey, 1970.
>
> J. DIEUDONNÉ, *Treatise on Analysis*, Vol. 2, Academic Press, New York, 1970.

Index

Approximation lemma, 37-51
  strongly continuous family of functions and, 45-50
Autonomous control problem, two-dimensional, 156
Autonomous equation, 4

Brouwer theorem, 128

Calculus of variations
  existence theorem for regular problems of, 162-170
  simple problem of, 163
Canonical systems of equations, 4-8
  Hamiltonian and, 5-6
Cantor function, 55
Carathéodory's lemma, 158
  proof of, 159-160
Chattering controls, 27, 51
Cone of variations, construction of, 120-129
Contraction mappings, fixed-point theorem for, 60-63
Controlled equations, 1
  unlimited extendability of solutions for, 146
  variations of, 99-106
Control parameter, 1
  set of admissible values for, 2
Control problem, 3
Controlled object, 2
Convex control problem, 24
  generalized controls and, 21-28
  varying of trajectories in, 99-113

Convex optimal problems, 24
  existence theorem for, 146-150
  maximum principle for, 120

Differential equations
  existence and continuous dependence theorem for solutions of, 53-77
  in general case, 72-76
  and homogeneous equation of variation, 86
  linear matrix, 93-98
  variation formula for solutions of, 79-98
Dirac measures, family of, 22-23

$E_{\text{Lip}}(G)$ spaces, 69-72
Equation of variation, 82-86
Equations, canonical systems of, 4-8
Equicontinuous family of functions, 113
Essential parameters, number of, 10
Existence and continuous dependence theorem, 53-77
  for solutions of differential equations in general case, 72-77
Existence theorem
  in class of ordinary controls, 150-156
  for convex optimal problems, 146-150
  for regular problems of calculus of variations, 162-170
Extremal motion, "kinematic" picture of, 11
Extremal solution, 9

Family of functions
  equicontinuous, 113
  smoothing, 39

Family of generalized controls, strongly continuous, 37
Family of mappings, 95-96
  equicontinuity of, 97
Family of operations, smoothing, 39
Filippov theorem, 135
Formula for the variation of the solution, 85
Function, majorant of, 88-89
Fundamental matrix, 94

Generalized controls, 21-35
  class of, 23
  convex control problem and, 21-28
  family of, 37
  maximum principle in, 118-120
  minimizing sequence of, 149
  as Radon measures, 140
  variations of, 99-106
  weak compactness of class of, 136-146
  weak convergence of, 28-36
  weakly continuous, 37

Hamiltonian, defined, 5-6
Hilbert–Tonelli existence theorem, 135
Homogeneous equation of variation, 84
Hyperplanes, family of, 86

Integral form, maximum principle in, 115-118
Integral maximum condition, 115-118
Integral uniform boundedness, 70-71

Linear matrix differential equations, solutions of, 93-98
Lipschitz condition, in differential equation solutions, 79
Lipschitzian functions, 54-55, 69-72
Lipschitzian set, 57-58, 70

Majorant, of function, 88-89
Mapping, nonsingular affine, 127
Mappings, family of, 95-97
Matrix, fundamental, 94
Maximum conditions
  in autonomous case, 12-15
  geometrical interpretation of, 11-12
Maximum principle
  in class of generalized controls, 118-120
  as first-order necessary condition, 19
  in integral form, 131
  proof of, 115-134

$n \times n$ continuous functions, 94
Nonsingular affine mapping, of simplex Λ into simplex X, 127

Open set U, 16-19
Operations, family of, 39
Optimal control, 3
Optimal control problems, canonical formalism for solution of, 16-19; see also Time-optimal problems
Optimal solutions, existence of, 135-170
Optimal transfer time, 3
Ordinary controls, existence theorem in class of, 150-156

Parameter(s)
  elimination of, 7
  essential, 10
Perturbations
  finite family of, 108
  of generalized control, 100
  sequence of, 129
Phase motion, 2
Phase point, 1-2
Phase velocity vector, 1, 161
Pontryagin maximum condition, 5, 115-118
Pontryagin maximum principle, 8
Probability measures, finite family of, 23

Radon measures
  convergence to, 141
  family of, 21, 145
  generalized controls as, 140
  mapping as, 142
  norm or total variation of, 31, 46
  strong convergence of, 32
  weak compactness of set in space of, 135-146
  weak convergence of, 28

Seminorm
  for arbitrary compact set, 80
  defined, 55-56
  truncated equation and, 91-92
Set, uniformly Lipschitzian, 57-58, 82
Sliding optimal regimes, 156-162
Sliding regime, uniform limit of trajectories and, 156
Sliding regime control, 27
Smoothing family of functions, 39
Smoothing family of operations, 39

Index

Solutions
   unlimited extendability of, 146
   variation of, 85

Time-optimal control, 3
Time-optimal problem(s); *see also* Optimal control problem
   with fixed end points, 3
   maximum condition in, 11-15
   open set U in, 16-19
   Pontryagin maximum principle and, 8-11
   statement of, 1-4
Trajectories, variations of, 107-113
Truncated equation, 91-92

Uniformly Lipschitzian set, 70, 81-82
Unity, partition of, 38-44

Unlimited extendability of solutions property, 146

Variation formula
   defined, 85
   for differential equation solutions, 82-86
Variation of the solution, defined, 84
Variations
   cone of, 120-129
   of generalized control, 100

Weak compactness, of class of generalized controls, 136-146
Weak convergence, of generalized controls, 28-36
Weakly measurable family, 21-22